A Fossicker's Guide to Gemstones in Australia

Nance and Ron Perry have been involved with stones, gems and gemcutting individually and together for more than forty years. They established a gemcutting academy and opal-cutting workshop which grew to include a lapidary supply company and an opal export agency. They have also been involved in the heat treatment and multiple machine cutting of Australian sapphires.

Both Nance and Ron have studied and gained the diploma in gemmology from the Gemmological Association of Australia and became Fellows. Their other books on gemstones include: *Australian Gemstones in Colour*, *Australian Opals in Colour*, *Gemstones in Australia*, and *Practical Gemcutting*.

Guide to

A Fossicker's
Gemstones
in Australia

Nance and Ron Perry

Authors' acknowledgments

Our grateful appreciation is extended to all who assisted us in gathering and upgrading the material for this book. We would particularly like to mention officers of the state departments concerned with mines, minerals, energy, resources and the environment; those dedicated people and friends involved with various airlines and tourist associations, particularly on the gem fields and in provincial centres; and last but by no means least, the postal staff who have been so helpful, free and generous with their time, advice and directions when we were struggling to find particular persons or sources of information.

We would thoroughly recommend that would-be fossickers consult all these fine and friendly people if and when they have a problem in unknown territory. We thank them all!

This edition first published in 1997 by Reed Books
a part of Reed Books Australia
35 Cotham Road, Kew, Victoria 3101
a division of Reed International Books Australia Pty Ltd

First published as *A Prospector's Guide to Gemstones* in 1982
Reprinted 1988

Copyright © Nance and Ron Perry 1982, 1997

All rights reserved. Without limiting the rights under copyright above, no part of this publication may be produced, stored in or introduced into a retrieval system, or transmitted in any form or by any means (electronic, mechanical, photocopying, recording or otherwise), without the prior written permission of both the copyright owner and the publisher.

National Library of Australia
 cataloguing-in-publication data:

Perry, Nance.
 Fossicker's guide to gemstones in Australia.

 Bibliography.
 Includes index.
 ISBN 0 7301 0500 8.

 1. Precious stones - Australia - Guidebooks. 2. Prospecting - Australia - Guidebooks. 3. Australia - Guidebooks.
 I. Perry, Ron. II. Title

622.1880994

Text design and page layout by R.T.J. Klinkhamer
Printed and bound in Australia by Griffin Paperbacks

Contents

Map list *vii*

Introduction *1*

Fossicking basics *3*
    geography, geology, gemmology, method

Fossicking practicalities *19*
    legalities, etiquette, safety

Working alluvial gemstone deposits *25*
    beach collecting

How to recognise gemstones in the field *33*

Australia—the fossicker's paradise *39*

All states guide to gemstone localities *45*

New South Wales *69*
    New England district, Lightning Ridge, White Cliffs, gemstone occurrences

Queensland 93

Agate Creek Fossicking Area (Forsayth), Big Bessie Fossicking Area (Sapphire), Glenalva Fossicking Area (Anakie), Graves Hill Fossicking Area (Sapphire), Middle Ridge Fossicking Area (Rubyvale), Tomahawk Creek Fossicking Area (Rubyvale), O'Briens Creek Fossicking Area (Mount Surprise), Yowah Fossicking Area (Cunnamulla), Chinchilla, gemstone occurrences

Victoria 111

Beechworth district, Buchan district, Heathcote, Castlemaine district, gemstone occurrences

South Australia 119

Andamooka opal fields, Coober Pedy opal fields, Mintabie opal fields, gemstone occurrences

Northern Territory 131

gemstone occurrences

Western Australia 139

gemstone occurrences

Tasmania 147

Flinders Island, gemstone occurrences

Glossary 153

Further reading 156

Helpful information 157

Index 159

Map list

Opal-bearing areas in Australia *40*
Principal gemstone areas of Australia *42*

New South Wales

Principal rivers and gemstone-fossicking areas *70*
Bathurst area *71*
Tamworth area *71*
Goulburn area *72*
Far north coast *72*
Western New England tableland *74*
New England district *75*
Glen Innes area *77*
Lightning Ridge opal fields *83*
Opal fields centred on Lightning Ridge *83*

Queensland

Principal river systems and potential opal-bearing areas *96*
Western opal-bearing areas *97*
Northern gem fields *100*
Central sapphire fields *101*
Central gem fields *103*
Southwestern opal-bearing areas *104*
Yowah opal fields *105*

Victoria

Principal rivers and gemstone-fossicking areas *112*
Gemstone areas *114*

South Australia

Principal river systems and potential opal-bearing areas *120*
Opal fields *121*
Andamooka opal fields *122*
Coober Pedy opal fields *126*

Northern Territory

Principal rivers and gemstone-fossicking areas *132*
Alice Springs – Harts Range area *133*

Western Australia

Principal rivers and fossicking centres *140*
Southwestern gem fields (official designation) *141*

Tasmania

Principal rivers and gemstone areas *146*

Introduction

THE IMAGE of the old-time fossicker-prospector is familiar to us all: the battered appearance, the gaggle of gear and the swag on the back are automatically recognised as the mark of the searcher after nature's treasures. The breed has sometimes been looked down upon in the past, but with a general increase in leisure time and the advent and availability of the four-wheel-drive, off-road vehicle, the pursuit of fossicking has been taken up by millions of enthusiasts worldwide, who not only have come greatly to respect the calling, but have made it their own. Not all fossickers are, have been or ever will be supremely successful, but the encouragement which moderate success provides serves to build enthusiasm and knowledge. This book strives to assist those with the fossicking urge, and we hope that it will stimulate at least some to sample the unfettered freedom of the outdoors.

Many years have passed since in 1935 the first group of people gathered together to foster their common interest in gemstone fossicking. Since this group's inception, interest in fosssicking has grown, multiplied and prospered to the point where fossickers make up one of the world's largest single-hobby communities.

With the passing of time the fossicking family has gradually aged and grown, so that it now includes not only the highly physically active young fossicker but more mature enthusiasts who pursue associated activities that are not so physically demanding.

Over the years we have been in touch with many thousands of fossickers in all parts of the world. Specimens found by our friends in distant parts such as Switzerland, the United States, India, Sri Lanka, the United Kingdom, France, Germany, Italy, Brazil, Japan and Russia, as well as New Zealand and all states of Australia, have been swapped or purchased, sometimes accompanied by slides, sound tapes or films which explain the circumstances in which they were found. On overseas visits the private and public geological collections and museums have been our fossicking grounds. Photographs or postcards that we have obtained have enabled us to study many specimens that are not readily available in our own country. The more knowledge attained the better fossicker one becomes.

Pen friends and members of organisations in the fossicking or lapidary fraternity have become long-term friends, and have meant that the warmth of the worldwide fossicking family is a really tangible experience for us.

The recent financially difficult years brought to many an awareness of the value and security of portable property in the shape of gems and gemstone rough. This awareness was reached by some groups of people many centuries ago, and today the amateur is in a position, to a degree, to follow their example, not only by purchasing gemstones but by finding the raw material and then by proceeding to learn how to cut and polish it into a valuable asset. This is so true that we know of instances where families have found not only pleasure and relief from financial pressures through fossicking, but business and economic independence.

Material success, of course, is not the inevitable consequence of fossicking, but the possibility is definitely there, provided an intelligent and diligent approach is made to the subject, coupled with a modicum of luck.

We wish all success in their endeavours.

Nance and Ron Perry

Fossicking basics

THE DICTIONARY DEFINITION of 'fossicking' is 'To search for by digging out, to rummage or hunt about, search, dig out, hunt up ... ' The dictionary adequately describes the business of fossicking as many will readily recognise. The process seems quite straightforward and simple, which it is provided we know firstly what we are searching for and secondly where to look.

Many thousand kilometres can be travelled, many tonnes of soil and rocks moved with much backbreaking toil in search of desirable gemstone to no avail, or at most with very small result, due to a lack of this essential knowledge. There are rules and requirements for successful fossicking, as in all worthwhile undertakings, and unless followed all that is achieved is much vigorous activity, very desirable from a keep-fit standpoint, but by no measure the object of the exercise. There are four main headings under which the rules of fossicking may be set. The first of these is geography, where do we go to look; the second is geology, what do we look for when we get there; the third gemmology, how do we know what we are looking at; and the fourth subject is method, how do we go about collecting what we find. A reasonable balance of knowledge about all four important subjects is essential for a chance of fossicking success.

A realisation that nature is methodical and has placed the gems of its creation in fairly definite places, with a reason for each location, will also assist one to grasp the full picture sooner rather than later.

Geography

When planning a fossicking expedition, distance to be travelled as well as climate should be considered. It is also crucial to take relevant maps, and to ensure that you have the necessary skills to interpret them.

Distance

Australia is a vast continent and its very widespread nature makes fossicking quite a travel-intensive operation. Sound transport is most necessary and excursions to more remote and inaccessible places, where roads may be ghostly tracks or almost imperceptible, call for the use of vehicles such as the four-wheel-drive, off-road type. Many keen fossickers would not hesitate to venture into this type of area, but the rules of fossicking must always be kept well in mind.

Juniors study their collection from a fossicking trip. The large jar contains many waterworn agates collected from Bellambi beach on the south coast of New South Wales.

Climate

The climatic conditions in outback areas are often very severe. High daytime temperatures and low night extremes have caused much hardship, both to those lost and to their would-be rescuers. Before embarking on an expedition it is important to contact public authorities and make sensible enquiry regarding climatic conditions. Sufficient informed preparation can then be made with confidence. Motoring organisations in all states are veritable fountains of information in this regard and will be found to be helpful in providing 'present condition' of roads and other facilities. Of course, all fossicking areas are not at the end of hazardous travel, but attention to all details of even the shortest trip will result in a far safer, satisfactory and happy result all round for every member of a group.

Maps

A reasonable knowledge of his or her whereabouts at any point in time is a must for the fossicker in order not to be in a constant state of 'being lost'. Some of us are blessed with an instinctive knowledge of where we are and which direction is the right one, but the majority of us less gifted souls must become moderately proficient at recognising the physical make-up of the landscape from those carefully prepared representations on paper called maps. Mapreading skills should be developed to the point where the ability to recognise salient points on the landscape from a map is almost second nature.

Maps are a very necessary part of any excursion planning. General locality maps are needed to reach your intended destination, while a detailed fossicking area map is useful to pinpoint the particular spot for attention. The fossicking map may be a composite of a small-scale road map and added information gathered from geological or 'previous-excursion maps'. The latter should be part of every fossicking diary; they should record experiences and information gathered on previous expeditions. Previous-excursion maps should show geological formations which are particularly recognisable as landmarks, water supplies, pick-up points for provisions, meeting places, road junctions, areas noted which show promise for fossicking but are not yet investigated, and any other information which could be of use.

Hand-drawn maps should be checked against, or used in conjunction with, official maps from government instrumentalities. Older maps, too, quite often yield helpful information; they may show features which have been omitted from more recent editions and which could be of value particularly for the fossicker searching for an old mine or something similar.

State and federal lands, mines and resources departments, as well as travel bureaux, have a large assortment of dependable maps and other papers available, usually for a small fee; these will supply much of the information required to plan an expedition satisfactorily. It is far better to have an oversupply of information on an excursion map than too little. In addition to maps and leaflets, some state departments have issued very fine publications with the particular aim in mind of assisting the fossicker to know what type of gemstones to look for in their state, and where these gemstones can be found. At time of writing, such booklets are available from all state departments, many of which are known as Departments of Mines, Energy, Resources, the Environment or combinations of similar titles. See the list of addresses on pages 20–1 of this book.

It is most necessary, and this point cannot be overstated, that full information be gathered and understood before embarking on a fossicking trip. Without proper planning the fossicking expedition will not have any reasonable chance of success.

FOSSICKING BASICS

Geology

The body or sphere which we call the Earth was once a mass of whirling gases, which over aeons of time cooled to the point where the Earth became solid but plastic. With a further passage of time the surface began to crust and continued to do so until a sufficient surface solidity eventuated, which contained the still-molten interior. This cooling process is still incomplete, a fact which is evidenced by the division of the zones of solidification into the outer crust, the continental mantle and the inner core (which is still in a molten or at least plastic state). The solid crust of the Earth, which acts as a restraining influence for the remainder of the sphere, is not a homogenous body but is constantly

Timor Rock in the Warrumbungle Mountains, New South Wales, is an example of a volcanic core.

An example of the force of wind erosion. This weathered face of rock on Uluru (Ayers Rock) in the Northern Territory is called 'The Brain'.

being changed by forces acting on it from within and without. Cracks in the surface allow penetration of water and dissolved gases; these substances in turn act on the material underneath. The reactions that occur alter the physical make-up of the Earth's surface. Regions of the Earth re-form; they may attain a state of comparative equilibrium, or alternatively the re-forming process may cause further changes and new interactions between chemicals. The process is a continuing one and will go on indefinitely.

Rocks

The original molten material of the Earth crusted on the surface to form a group of rocks described as *igneous* (from fire), or *primary*. The atmosphere, in conjunction with water, reduced these primary rocks by erosion, and the debris of this erosion settled in areas where it was distributed by the forming rivers or by wind. Accumulations of these sediments, solidified by pressure from above or by the cementing action of other mineral solutions passing through them, formed the *sedimentary* rocks, or *secondary* rocks. Sediments were carrried by rivers and in the process the particles were separated according to size, with the larger, heavier ones being deposited upriver and the finer ones often being carried downstream to lakes or the sea. The saltiness of very large bodies of water often reflects the nature of the terrain through and over which the feedwater passes.

The clay, sand and gravel we have come to know are the size divisions of eroded debris produced by this process.

The carriage of sediments from one place to another causes a great change in the loading on the Earth's crust. A mountain may be literally moved from one place to another—in the form of eroded sediment—and, as will be readily visualised, there will be a corresponding transfer of the great weight of that mountain to a new location. We may think of the crust of the Earth as being solid, but this is purely relative. Faults, fissures, cracks and other unstable features exist beneath our feet despite appearances to the contrary. With an unbalanced new loading on any area, something has to give, and so it does. Earthquakes are just the Earth readjusting the balance—moved mountains have a similar effect. The movement of the Earth's crust that occurs during an earthquake puts great stress upon existing rocks with changes in their form quite often the result. This changing, or metamorphosis by heat and/or pressure produces another type of rock which is called, naturally enough, *metamorphic*, or *tertiary* rock. The heat and pressure needed to cause such transformation does not necessarily result only from Earth movement; it may also result when new molten lava or magma from beneath the crust intrudes into fissures, cavities and cracks produced in the existing strata by Earth movement. When the molten lava or magma in this latter process cools *intrusive* rocks are formed. The action of heat and pressure on existing rocks will often change their composition, as the molten intrusion may contain constituents which are different from those of the rocks and which interact with the rocks to form new materials. Such changes can often form minerals which are considered by man to be desirable—gemstones.

From the foregoing, it will be seen that the formation of gemstone minerals is a very miraculous and wonderful process, but is still an orderly and predictable one. The only unpredictable factor is, where *exactly* is the fossicker to dig! It is most necessary to study the area to be fossicked from a geological standpoint. To attack a great mass of solid rock to extract its treasure of gems is sheer madness. To quietly search out a cavity in a rock formation that suggests the presence of gemstone is much wiser.

Minerals

Confusion over what comprises a mineral and what a rock can be easily overcome if the rocks are remembered as the building blocks of our planet while the minerals are the constituents of those blocks. A rock may consist of a single mineral or a mixture of minerals. For example, quartz is a mineral made up of silicon and oxygen; granite is a rock made up of a mixture of the minerals quartz, mica and feldspar.

Gemstones found in the various rock groups

The igneous rock group produces many gem materials. Among them are the *granitic* rocks feldspar, quartz, zircon, topaz, tourmaline and sphene, as well as gems formed by *contact* or *regional metamorphism*—emerald, beryl, chrysoberyl, feldspar, garnet, kyanite, sapphire, ruby, spinel and spodumene. (Contact metamorphism is the result of heat while regional is the result of pressure.)

FOSSICKING BASICS

Prase deposits at Hanging Rock, Nundle, New South Wales. The rock is metamorphic.

Columnar basalt at Mount Hoy in north Queensland. This is a form of cooled lava.

Pegmatites are the rocks formed by the interaction of magma intruded into existing rock cavities, and are regarded as good prospects for gem-type minerals. The minerals formed will depend, of course, on what type of magma is squeezed into what particular rock cavity and the composition of both. It can be imagined how the molten material would tend to dissolve some of the cold rock, while at the same time the cold rock would have a cooling influence on the hot magma. This countereffect of one upon the other accounts for the variety of crystal sizes found in a suitable cavity, the smaller ones having been formed initially when there was rapid cooling and the larger ones as the whole body became uniform in temperature and proceeded to cool more slowly. The slower the cooling, the longer the time for formation and growth and the larger the ultimate crystal.

The formation of the so-called 'hydrothermal precipitation of crystals in nature' occurs where the solution

An attractive agate geode containing crystals of purple quartz (amethyst).

FOSSICKING BASICS

Smoky quartz with tourmaline inclusions from Mount Isa, Queensland. The small crystals of tourmaline can be seen emerging from the outer skin of the smoky quartz.

derived from water and the minerals contained in rocks attain suitable conditions of concentration and temperature. A crystal will grow if the solution containing the necessary constituents reaches a concentration that cannot contain more of the mineral. Crystals of each mineral will grow independently of one another. Dissimilar minerals may grow one within the other—for example, tourmaline in quartz—when occurring in the same mother liquor in solution. Quartz, in all its varieties, is the most common of all hydrothermally formed gem minerals, such varieties being created over a wide range of temperatures. Opal, a specific variety of quartz, occurs only at the lower end of the temperature range, being non-crystalline in nature. The upper range high-temperature minerals are essentially crystalline and include garnet, tourmaline and topaz. The lower range of non-quartz, high-temperature minerals includes rhodochrosite and calcite. Agate is one form of hydrothermal quartz—contemplation of a sliced agate geode leads one to realise that the solutions which penetrated the cavity to form the deposit must have been under considerable pressure. Too often the enormous forces of nature required to produce just one beautiful, perfect agate geode are not appreciated by the inexperienced fossicker.

Erosion of the surfaces of the Earth's rocks causes the release of many gemstone types. Erosive forces assist the fossicker, not only by wresting the gemstones from their inaccessible place of formation, but by concentrating them, in many instances, for more or less casual collection by those who learn where to look. Sapphire or garnets are released from their mother rock, and geodes and cavities containing the quartz varieties amethyst, citrine, agate, chalcedony, et cetera are freed from their beds of primary or secondary rock. All may be carried away by rivers or streams to be selectively deposited in wash which may be covered up by new sediment or caught up in further upheavals of the Earth's crust and thus incorporated into new formations. The sapphire deposits of northern New South Wales and central Queensland are typical examples of the effects of erosion, concentration, relocation and re-formation. The 'deep leads' which contain waterworn and/or fragmented crystals are veritable treasure houses for the lucky fossicker.

The passage of time has, in some instances, stacked these deep leads one above the other as rivers changed course and relocated, only to change direction and flow again, this time possibly immediately above the first bed. Core drilling on the Anakie gem field has revealed evidence of deep leads set one above the other in excess of six layers in some areas.

A realisation of the large variety of gemstone deposit types comes very gradually to the new fossicker, but as it comes a great realm of possibility unfolds. Deposits of quartz gemstones, for example, are not only available as massive lumps of colourful jasper, but exist also as much smaller, portable pieces to be found in rivers and streams as well as on beaches and in dry places far from any evident water. All these deposits are the result of the action of the forces of nature, formation, erosion and distribution, and all these processes work to assist the fossicker. Faulted gemstone pieces, when carried by water, are reduced to sound smaller pieces which will suit the fossicker's purpose. Heavier or larger gem pieces are deposited high in streams while smaller ones travel further downstream where the fossicker can decide which will suit his particular wants.

Many gemstones are concentrated by other natural means. Freezing temperatures, acting on water contained in fissures in a rock's surface, will cause expansion which will in turn separate flakes from the surface of the rock and ultimately release the constituent minerals. The coarse gravels which result often contain useful and desirable collectors' pieces. This process is very evident in granite.

Gemmology

A knowledgeable answer to the question 'What gem are we looking for?' comes only after a deal of serious study of the fundamentals of gemmology and/or mineralogy. The latter is the study of all minerals, while the former limits the study to the smaller number of minerals which are gemstone forming. A handbook, such as our book *Gemstones in Australia* (Reed, 1979), will provide adequate information on identification of the usual gemstones encountered by the fossicker, but if rare or out-of-the-ordinary types are encountered further reading may be necessary.

Gemmology is a vast subject which, no doubt, the dedicated fossicker will investigate much more deeply than could be dealt with in this book. We recommend that fossickers enlarge their knowledge in this area as early as possible in order to increase the enjoyment of the hobby.

Method

The collecting of gemstones can be as simple as bending down and picking up a small piece from a beach. Or it can be as arduous as swinging a heavily laden sieve or spending the day at the end of a pick and shovel or sledge hammer. Nature has distributed her treasures widely and has put much dirt and rubbish between the finer pieces of her creation. A good adage for the fossicker who has come upon a piece of promising gemstone is not, as some seem to think, 'if you can't see what it is, hit it'. Many a fine gemstone piece has been ruined by such action. Remember always that old motto told by a mineralogist, 'Nature took millions of years to make that piece, so treat it with reverence'.

Aids to the human hand are necessary for fossicking. The most essential tools needed are the geologist's hammer and a bucket. Add to these a small short-handled shovel, easy to use in confined places—a collapsible trench shovel is favoured by many. Also necessary is a long-handled, round-nosed shovel for moving large amounts of dirt. And a small, short-handled miner's pick and pick mattock are helpful if large excavations are contemplated.

A one-kilogram lump hammer and a sledge hammer of suitable weight for the user are good driving tools. A selection of cold chisels from 150 millimetres to 200 millimetres long plus a few rock gads are necessary for penetrating cracks and splits in rocks. Wedges are also needed for this purpose. A collection of gathering tools will be needed but these can quite readily be improvised from the usual gardening implements found around the home.

A three-pronged scratcher, a small fork of the hand variety and a narrow putty knife all will be found useful for scratching around in gravels and broken rock. A pair of stout, short, leather gloves such as are used by steel handlers are a safeguard against mashed fingers when using the hammers. They also help ward off any aggressive noxious insects which might cohabit with desirable crystals in cavities! A bent-ended screwdriver or two of large and smaller dimension will be useful for scraping out these cavities. For recovery of alluvial gemstones and minerals, a pair of sieves, one with a three-millimetre and one with a six-millimetre mesh, together with a similar-sized gold pan, a heavy hessian bag, a pair of tweezers and a hand lens will make a base kit. Any specialised tools which the fossicker

FOSSICKING BASICS 17

may find good for a particular job are always added as part of essential equipment, and some of the most useful tools are those dreamed up to meet an emergency.

The fossicker's diary referred to earlier is just as essential a tool as any, and fossickers should take the time to record the details of each expedition. Whatever implements you have at your disposal, it's no use digging if you don't know where to dig!

A selection of hand tools is illustrated which the fossicker will find of great use. The Hunter II, and the Mini Hunter are very useful for field and beach use where foot-assisted digging is employed to find buried gemstones. The Wide Blade Trowel is ideal for probing and shuffling loose gravels into sieves and dishes from potholes and crannies where a longer handle would get in the way.

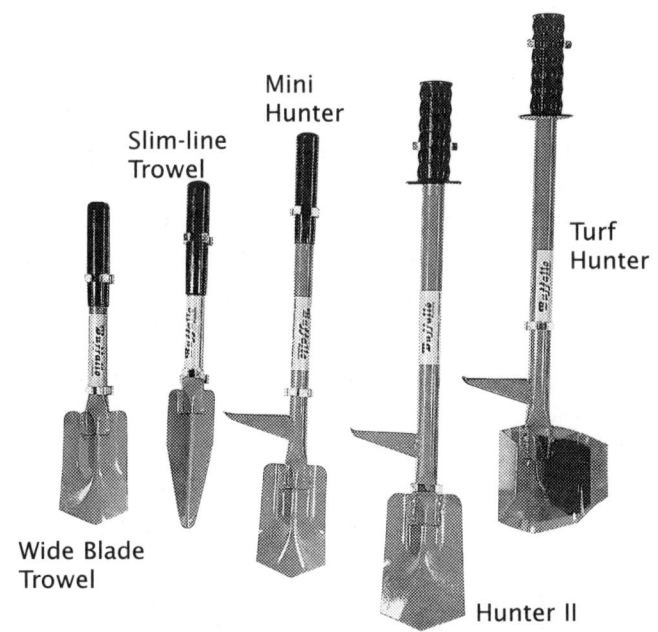

Fossicking tools. A selection of digging implements especially designed for the gemstone fossicker.

These specialised tools have been designed with the fossicker/treasure hunter in mind. They are advertised throughout Australia in the various lapidary, gemmological and treasure hunting journals and should be available from most rock shops.

Fossicking practicalities

Legalities

IT MUST BE REMEMBERED by the fossicker that although a particular gemstone has at some time been reported as found in a certain area, there is no guarantee that more of the stone will be found there. Then again, as has been proved on abandoned opal and sapphire claims, valuable finds could as yet be undiscovered. Gem fossicking is a game of chance, where the only certain rewards will be the enjoyment of the wide open spaces and the country travelled, the companionship of friends and family who are engaged in a common pursuit, and the thrill obtained from finding *some* type of stone for your collection, even if only a piece of jasper, topaz or chalcedony, which in themselves can be quite beautiful.

In this world of change nothing is permanent, not even fees for mining or fossicking licences. We will not list here the fees charged in each state, as by publication date there may be some variation. It is sufficient to say that fees, if there are any, are not exorbitant and vary from state to state. Each authority has its own scheme of permits and regulations, and even the departmental naming is widely different in each state. However, in these days of extended tourist promotions and increased interest by government departments and agencies in such activities great efforts have been made to provide assistance to fossickers, whatever their interest. All states issue comprehensive information in brochure form, as well as maps, which are in some instances very detailed indeed.

Before embarking on a fossicking expedition it's very necessary to obtain the appropriate documentation from the state authorities and to observe the list of regulations therein governing fossicking and mining activity. The appropriate department should also be able to supply information on gem areas which may be available. Here are the addresses for the mines departments in each Australian state:

Department of Minerals and Resources
PO Box 536
St Leonards
NSW 2065
Phone (02) 9901 8888

Queensland Minerals and Energy Centre
GPO Box 2564
Brisbane
Qld 4001
Phone (07) 3237 1659

Department of Natural Resources and Environment
PO Box 41
East Melbourne
Vic. 3002
Phone (03) 9651 7011

Department of Mines and Energy
191 Green Hill Road
Parkside
SA 5063
Phone (08) 274 7500

Northern Territory Department of Mines and Energy
GPO Box 2901
Darwin
NT 0801
Phone (08) 8999 5286

Department of Minerals and Energy
100 Plain Street
East Perth
WA 6004
Phone (09) 222 3333

Department of Development and Resources
PO Box 56
Rosny Park
Tas. 7018
Phone (03) 6233 8333

It's also recommended to contact the local mining warden in the district you visit. Intending fossickers should take heed of the fossicking etiquette and safety considerations outlined below. All this should provide an almost sure-fire guarantee of a successful expedition.

Even though a person may hold a licence, there are still certain restrictions imposed upon him or her. These may vary from state to state, and, as has already been mentioned, it is always a good plan to contact the appropriate government department or the local mining warden. As a general rule:
- The fossicker may only use manually operated tools, and must not cause soil erosion or pollution.
- The fossicker must not excavate more than two metres in depth, and all excavations must be filled in when no longer required.

- A limit of ten kilograms is placed on the amount of material which can be removed in any period of forty-eight hours.
- Conditions of entry to private land declared as a fossicking area may be imposed by the owner.
- Certain areas of crown land may differ from place to place in their fossicking requirements.

Etiquette

It is required of the fossicker that he or she always respect the rights of property owners, and obtain their permission before entering land owned by them.
- Leave gates as found, whether open or closed.
- If permitted to light fires, always put them out when leaving.
- Leave some stones for the next guy.
- Bury all rubbish, and fill in all pits and holes dug while fossicking.
- Do not walk or drive across cultivated land.
- Do not carry firearms or disturb livestock in any way.
- Watch out for other people's claims and keep clear of them.
- Avoid 'prohibited areas' such as military reserves, Aboriginal lands and national parks.

Safety

As most fossicking areas are to a large extent 'off the beaten track' your vehicle must be in sound condition. Automobile clubs in each state issue brochures with hints on motoring in remote areas, as well as the necessary maps.

Aids to safety and comfort could include:
- maps
- spare parts for your vehicle
- cool and warm clothing, good walking shoes, hat with wide brim

- sufficient food, water and fuel for a longer journey than you anticipate
- sunglasses, insect repellent, lip salve.

When travelling outback, always inform someone of your intended destination and time of arrival. This will ensure that in a case of breakdown help may be alerted.

In these days of superior electronic and other aids for outdoor activity lovers, mention must be made of some simple common-sense safety precautions, which are necessary and in some instances required in observation of the law. These include the carriage of fuel in containers designed expressly for the purpose, the storage and carriage of drinking water in remote parts, again in the correct type of container, and last but by no means least the maintenance of liquid petroleum gas containers. Time speeds by when one is having fun, and years may pass by without testing your gas container. But note that the law requires that all gas containers be tested every ten years. Testing must be carried out at a licensed testing station for a fee—the various components subject to ageing are automatically replaced. The cost for the service varies according to the cylinder size. Other essential containers should also be inspected from time to time to ensure that they are sound and their markings clear and in good order; this is important to avoid the accidents that could take place from filling a container with the wrong substance. A diesel/petrol mix-up could be disastrous, as could a mix-up involving a drinking water container. Tents, sleeping bags, cots and mattresses are most necessary items for repair and maintenance before departure. To have the tent poles or pegs missing on arrival in the dark is not to be contemplated. A stuck or corroded zipper in a sleeping bag, an unserviceable groundsheet or a leaking air mattress are all unwanted problems. Lamps, torches and the like should all be tested as part of the overall safety preparation.

A more recently developed aid for use in the off-road vehicle is the Sony Satellite Navigation Positioning Unit presently much used by the boating fraternity. This very useful piece of equipment uses satellite information to provide accurate location; it would be invaluable for the fossicking group who are following the lead of a 'reported' gemstone occurrence in a remote area.

Working alluvial gemstone deposits

DEPOSITS OF GEMSTONES were formed in rocks many millions of years ago. Some of these rocks eroded and the gemstones then travelled from their birthplaces by means of rivers and other bodies of water to their present positions. As was noted earlier, the gemstones have been concentrated and the lesser and softer material which was carried with them has degenerated to silt. The larger and heavier pieces of stone will have been caught in pockets and holes higher upstream while the smaller pieces of heavy material and larger ones of the lighter varieties will have travelled further downstream. These will have been deposited selectively, depending on the velocity of the water flow and the depth of the water. They may have been deposited on the stream bed or in a pocket or hole, in front of or behind any obstruction which causes a lowering of the water speed to below that of the water which has carried them thus far. The widening of a river into a slower-moving body of water as it passes from a mountainside to the plains below is one example of a concentrating influence. The falling of a waterfall into a waterhole is another. The deep wearing effect of the water makes waterholes very good fossicking spots. Wetsuit and mask, at least, are usually necessary to investigate these spots as considerable depth and low temperature are characteristic of many such gemstone pockets.

A change of direction by a river will cause an accumulation of pieces which are too heavy to negotiate the new direction on the outer side of the bend, while the slowing of the current on the inner side of the bend will cause still smaller pieces to be deposited. Both locations are worthy of investigation. All deposits by water will be in the form of gravels, or *wash*. Wash can be just a few grains at random or many metres thick depending on the age of the river and the extent of the source of the mineral, its state of erosion and so forth. A little probing around with a sharpened piece of ten-millimetre reinforcing steel rod with a bent-end handle will be a help in locating a different-sounding wash which could be a bonanza. Wash is also found as discrete pockets rather than as a sheet formation, so it is advisable to investigate all irregularities in the line of a creek or riverbank for isolated deposits. The piece of rock jutting out into the water could have a deposit at its foot, both up and downstream. Rocky-bottomed waterways or shallow rock shelves over which water runs or has run for centuries can often be worn where cracks or other vulnerable spots occur in the make-up of the rock. These are prime places for fossicking. Scoops and cutlery-type spoons with bent handles often give access to difficult-to-reach potholes and crevices where smaller gem pieces settle.

The basic sieve combination of a six-millimetre mesh set on a three, possibly with a gold pan under it, is loaded with wash from the pocket and washed through until all material is separated. Place a wet hessian bag on a flat area and upend the coarser sieve on it in one movement. The contents will spread out in a reasonably even layer if the washing has been carried out in a regular 'dunk, twist and lift' action (this causes the contents of the sieve to move in a circular manner which, with practice, will concentrate the heavier pieces towards the centre of the whole). Look across the stones, and with the tweezers pick out any glistening pieces for examination. With a little practice you will soon be able to see which are just wet junk and which really have the desirable lustre of a sapphire, a piece of topaz or some other precious stone. Repeat the process with the finer sieve. Place any finds into a container and repeat. (Note that gemstone pieces look larger if placed in a bottle of water.)

This recovery operation can be backbreaking if attacked with too much vigour—particularly if you haven't done it before. A moderately paced 'half fill the sieve and wash' procedure should be developed, and provided the sieve isn't one of the older wooden-rimmed 'sixteen-incher' varieties this can be kept up for hours. The larger sieves hold a large weight of gravel and can leave the operator very short on energy at the end of the day. Keep some energy for filling in and cleaning up the digging site before retiring to camp to inspect the day's take. If a lot of cleaning is left until the end of a weekend dig, a deal of time will appear to have been wasted just filling in holes. It's better to clean-up and fill holes as you go. It is most necessary to fill fossicking holes both to avoid visual pollution and to avoid causing destructive erosion of the landscape. The filling-in procedure is even more essential when digging on private property; if the landholder has been obliging enough to allow you onto his or her property, it is decent to ensure that no animal traps be left in the shape of mini wells 'where that beautiful blue sapphire was found'. Remember also to take all cans and garbage when leaving, or to bury it securely in one of your holes. Always leave the place in a little better condition than when you arrived—if possible.

Alluvial deposits are not always found close to water. A watercourse could have deviated many kilometres from its original ancient course, leaving the debris brought down over centuries, or longer, as evidence of its previous path.

This debris, in turn, can become covered by other sediments, blown dust or deposits formed by radical Earth movements. The overgrowth of trees and other vegetation makes this type of deposit a little harder to find; it may be necessary to carry out a particularly thorough study of the fossicking area. The extraction process is also a little more difficult. Location of buried wash is achieved by digging potholes in the prospective area after going over it with the probing rod to find a likely starting point. Once a pothole has reached a show of wash, dig others a couple of metres away in several different directions.

Quite often this kind of deposit is located in wetland areas. Reedy types of overgrowth may cover a fairly high water table, and if one of the potholes is dug down below the top of the wash level, the hole will fill with water which can be used for sieving. Once wash has been located in more than one exploratory hole, the top layer can be stripped off to expose the gravel. This can then be treated in the same way as wash from a river.

Dry areas without any sub-surface water will make it necessary for the wash to be taken to a source of washing water. Should this be impossible, there are two alternatives open to the fossicker. The first of these is dry sieving which will serve to concentrate the wash into sizes suitable to take home or to a water source; the second method is to carry water onto the site. A great deal of water is not necessary. A halved 200-litre drum makes an ideal washing tank. A smaller container could be used but the lack of a settlement area will limit its use before cleaning out has to take place. Assuming that a half drum is used, filled two thirds with carried water, it is necessary to first sieve through the wash to separate the rubbish from the potential gem-bearing gravel. The rough dug wash is tipped through the first sieve and then the resulting gravel is placed into the fine sieve. This, in turn, is suspended over the drum of water by a cradle with a handle arranged so that the sieve and its contents can be jigged up and down in the water. The miners on the Anakie gem fields call this arrangement a *willoughbee*, or *willobie*. With practice a rhythm can be achieved that will concentrate the heavier material in the same way as in hand working. Once the sieve has been lifted out and upended on the bag, selection with the tweezers goes on as described earlier.

Concentrate from dry sieving is rather deceptive as far as weight goes. A small bag of gravel, particularly if it contains a proportion of gems such as garnet, sapphire, zircon and quartz, can be considerably heavier than sand alone. A vehicle used on a fossicking trip could be overloaded by the addition of more than a few bags of concentrate. Due consideration must be given to this matter when planning a trip.

Beach collecting

Gemstones washed up on ocean beaches are a very attractive proposition for fossickers as there is little or no digging required for recovery, as a rule. Location and selection are the two most important operations. Once the prospective area has been pinpointed and located, keen eyes and a strong back are the main tools necessary, combined with a receptacle for the finds. A long-handled scoop to reduce back bending has been seen in action, and after a day's collecting some might agree with its merits.

A knowledge of the probable gemstones to be encountered in an area is a great advantage, but in its absence a knowledge of the appearance of the waterworn rough in the silica family is recommended. Agate, petrified wood, jasper and chalcedony are the most frequent finds on beaches. These have been eroded from volcanic cliff faces by the oceans, and worn and reduced to smooth-surfaced pebbles of every size. Very large pieces of petrified wood have been washed up on the southern beaches of New South Wales. These showed rather less wear than would be expected; smaller pieces have shown far greater degrees of wear. One piece of petrified wood in our possession comes from Belambi Beach. It weighs only about 400 grams and is from a very large tree; in its travels it has been worn to a smooth kidney shape.

Small agate geodes that have very clean surfaces show the high degree of resistance to wear of the quartz gemstones. Geodes that show wear to the point of being open indicate the very long time that they have been in the sea. Jasper is a very common seashore alluvial gemstone which can be found in all colours as pebbles. These are very good stones for the tumbler polishing process.

The seashore fossicker has one very essential consideration to be aware of and that is degree of excellence.

An ocean beach at Port Macquarie, New South Wales. A good fossicking spot for petrified wood, agate and jasper, with possibilities of rhodonite.

On arrival at a new beach, everything is of interest. Every find is a good one and into the bucket it goes. After an hour or so, the bucket can become very heavy. It is a wise programme to have a coffee break, and while relaxing to empty the morning's collection onto a groundsheet and selectively sort it. Great care should be taken to reject any but the best and most desirable stones—in the eyes of the collector. If every piece is to be taken home, the fossicker will eventually possess an aggregation of 'junque', unusable 'junque' at that, which is not only unusable, but fit only for the cement mixer. Fossickers come in all stages of experience, and as their experience grows and enlarges gathered pieces will improve in quality and excellence.

Every piece collected should be sound and as free as possible from defects. Of course, the type of material collected depends on the ultimate use, but for polishing purposes the piece must be of a high standard. The surface of any unidentified stone must be smooth and hard if it is to be kept. Porous pieces of sandstone which dry readily when wiped with the palm of the hand or a cloth are unsuitable for polishing. Any stone which shows deep pits and scars should also be discarded.

How to recognise gemstones in the field

THE MAIN GUIDES to recognition of gemstones on a fossicking expedition are colour, size, and lustre (that is, whether the stone has a glassy or waxy appearance).

actinolite Rare; dull green; small needle-like crystals; usually occurs as inclusions in other minerals.
agate Grey, yellow, red, blue, white, black and brown. Roughly oval nodules, with banded or patterned interior. A variety of quartz.
alunite Pink, red, white, grey. Massive, with sweet taste of alum. Silky. An aluminium mineral.
amazonite Blue, white, banded. Compact form or tabular crystals. Silky appearance. A microcline feldspar.
amethyst Mauve, violet, purple. Crystal form. Sometimes contained inside agate geodes and thunder eggs. A variety of quartz.
andalusite Mixed or single colour: green, grey, yellow, brown, pink, red-violet or white. Granular or massive.

apatite Light yellow to yellow-brown, blue, green, black or colourless. Prismatic or granular; also massive or tabular crystals.

aquamarine Pale blue to blue-green; large crystals. One of the beryl group.

aragonite White to creamy white; short to long crystals; columnar aggregates. A calcium mineral.

australite (tektite) Black to green-brown. Glassy; obsidian-like; usually button- or dumbbell-shaped.

axinite Grey, brown, green; axe-shaped crystals. Rare.

azurite Blue. Usually found with malachite, in single crystals or aggregated. A copper mineral.

beryl Violet, white, colourless, pink, green, blue; can be transparent or opaque.

cairngorm Smoky quartz with orange shading.

carnelian Bright orange-red; waxy lustre; usually small pebbles. A variety of quartz.

cassiterite Red-brown, yellow, black; metallic. Rarely occurs without colour. A tin ore.

chalcedony Colourless, white, cream, green, pink, red, blue, brown; waxy lustre. A form of quartz.

chert Dense, opaque variety of quartz, with splintery to conchoidal fracture. White, grey, black, green, pink, blue, brown, yellow and red.

chrysoberyl Green, yellow, yellow-brown, blue-green. Botryoidal, massive, or crystal clusters.

chrysocolla Blue, blue-green to black. Soft. A copper ore.

chrysocolla in quartz Chrysocolla impregnated with quartz, with corresponding hardness. A copper mineral.

chrysoprase White to pale green to deep green. Waxy. Massive, in sheet-like pieces three millimetres to seventy-five millimetres thick. Also nodular. A variety of quartz.

citrine Golden yellow; crystalline, with conchoidal fracture. Small to large crystals, singly or in clusters. A variety of quartz.

common opal Mixed colours: yellow, brown, black-brown, red, white or green. Massive or as replacement of wood.

crocoite Red-orange; delicate crystal formation. Rare. A lead mineral.

diamond Small crystals; oily appearance. Pure carbon in crystal form.

emerald Bright to deep green beryl.

epidote Yellow-green to yellow-black.

fluorite Green, purple, brown, black and pastel shades. Crystals. A fluorine mineral.

garnet Red to red-violet, yellow-brown. Glassy; small to large crystals.

goshenite Colourless beryl.

heliodor Violet, gold or yellow beryl.

hematite Black. Iron ore mineral.

horn Cream, black, brown, grey. Greasy; of animal origin.

hyalite White, water-white form of common opal. Botryoidal. A silicon mineral.

iolite Blue, smoky. Shows two or more colours.

jasper Pink, yellow, orange, red, green, brown, black. Small to large pieces. An opaque variety of quartz.

jaspilite Bands of jasper and hematite. Red-brown with straight or wavy banding. Quartz in combination with iron.

jet Black; fairly pure form of lignite.

kyanite Blue; crystal form.

labradorite Yellow; transparent, with distinct cleavage. A plagioclase feldspar.

lazulite Blue to blue-green, white. Massive.

malachite Dark and light green patterns. Soft; massive. A copper mineral.

marcasite A crystallographic variation of pyrite (iron sulphide); soft.

mookaite A variety of chert.

moonstone White, clear, pink, grey. Pearly. Variety of orthoclase feldspar.

morganite Pink form of beryl.

morion Dark form of smoky quartz.

nephrite Green, black. Tough, loose, weathered boulders.

obsidian Black or brown, transparent to translucent volcanic glass.

opal Rainbow colours. Freestanding, or on grey or black potch. Queensland opal in ironstone, or in 'nuts'. A hydrated form of silica.

opalised wood Typical wood grain. Wood colours: cream to brown. Conchoidal fracture. A fossil that has been replaced with quartz.

opalite Yellow, blue, cream; usually with dendritic markings.

'peanut' jasper Brecciated form of jasper; a cemented quartz conglomerate.

peridot Yellow to green; small stones.

prase Pale green to brown; dense, opaque quartz.

prehnite Yellow-green; botryoidal masses of radiating fibrous crystals in basalt and diorite.

pseudomalachite Similar colour and form to malachite but harder. Rare.

pseudophite Green to black type of serpentine.

psilomelane A heavy, mottled black, massive manganese mineral.

pyrite Brassy, metallic crystals in striated cube and massive form. A sulphide of iron, sometimes mistaken for gold ('new chum gold', or 'fool's gold').

rhodochrosite Watermelon pink to red; sometimes banded with white. Usually massive; sometimes in crystal clusters. A manganese carbonate.

rhodonite Pink, red, black; usually massive. A manganese mineral.

ribbonstone   Cream, pink, red, yellow, orange, brown patterns. Large pieces. A form of chert.
rock crystal   Transparent crystals, often in clusters. Variety of quartz.
rose quartz   Pink form of quartz.
ruby   Red, sometimes with mauve shading. Small stones. A variety of corundum.
rutilated quartz   Clear quartz with inclusions of tufts of fine needles of rutile. Red-orange, silver, golden or brown shades.
sapphire   Bright appearance; blue, green or yellow. Waterworn or crystalline.
smoky quartz   Brown, transparent form of quartz; usually in crystals.
sphene   Yellow, green-brown and grey. Small.
spinel   Yellow, green, red, brown, black crystals.
steatite   Also known as soapstone; a variety of talc. Greasy-feeling massive material; may range in colour from white through cream, with or without mottling of other colours.
stichtite   Red to lilac with flecks of a darker colour dispersed throughout; soft; massive in form. May be scratched with the fingernail. So far found only in Tasmania.
tektite   See *australite* above.
thunder egg   Star-shaped agate or amethyst filling in spherical nodule (often large).
tigereye   Red-brown, yellow-brown, blue, red. A silicified asbestos.
topaz   Blue, pink, colourless; small to large crystals with striations parallel to their length.
tourmaline   Green, pink, black; in crystals, with deep striations parallel to the length of the crystal.
tourmaline in quartz   Colourless or smoky quartz, with inclusions of black tourmaline.

turquoise Pale blue to pale green; often in matrix. Aluminium phosphate with copper and iron.

variscite Light green to emerald green; massive; sometimes brecciated.

zebra stone Striped variety of siltstone.

zircon Brownish red, square prisms terminated by pyramids; massive; clear; small. Glittery appearance. A zirconium mineral.

Australia— the fossicker's paradise

AUSTRALIA IS SITUATED between 9°20' and 43°39' south latitude and 112°50' and 153°45' east longitude. The country's total area encompasses 7 682 300 square kilometres, which comprises approximately five to seven per cent of the total area of the Earth excluding Antarctica. The length of the coastline, including that of Tasmania, is 36 735 kilometres. The distance from Cape York in the north of the country to South-East Cape in Tasmania is 3680 kilometres as the crow flies; from Dirk Hartog Island in Western Australia to Cape Byron in northern New South Wales the distance is 4000 kilometres. This very extensive island continent is sparsely populated, most settlement being on the eastern and southeastern seaboard.

The land mass of Australia has undergone great geological changes over the hundreds of centuries since its formation, and it has been altered further by the influence of climate. There have been movements of whole geological structures from place to place, in total or in part, by the influence of wind, water, heat, cold and/or direct upheavals of the Earth such as earthquakes or volcanic action. Large areas which were once submerged have been elevated; land that was once dry has been submerged. Land has been elevated, depressed, buckled, twisted, relocated, compressed, faulted and moved to convert plain to mountain, mountain to valley or to level entire mountain chains. For example, a very large area which today is comprised of parts of Queensland, New South Wales, the Northern Territory, South Australia and Western Australia was in ancient times submerged by a vast inland sea which has since dried up. This area now largely comprises what is termed the Great Artesian Basin, a sub-surface water-bearing area.

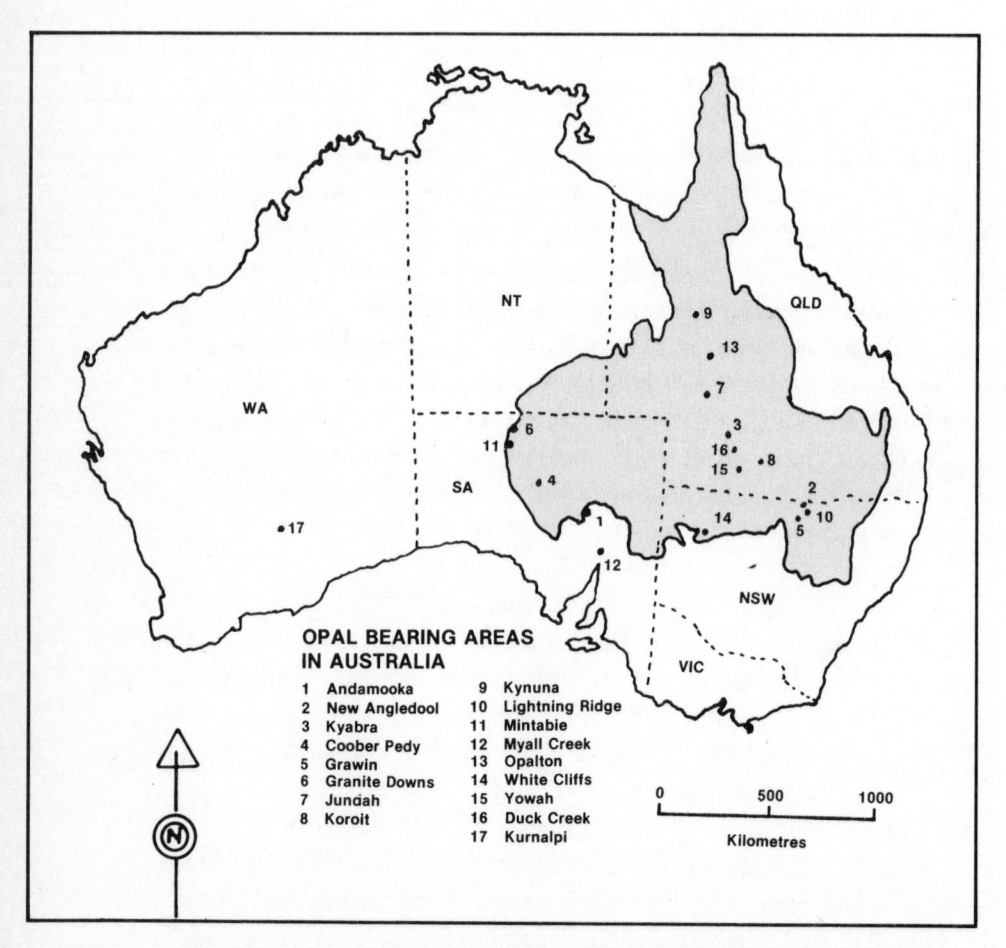

OPAL BEARING AREAS IN AUSTRALIA

1. Andamooka
2. New Angledool
3. Kyabra
4. Coober Pedy
5. Grawin
6. Granite Downs
7. Jundah
8. Koroit
9. Kynuna
10. Lightning Ridge
11. Mintabie
12. Myall Creek
13. Opalton
14. White Cliffs
15. Yowah
16. Duck Creek
17. Kurnalpi

Opal-bearing areas in Australia. (Information by courtesy of the South Australian Department of Mines and Energy.)

Underground water is still present in this region, and in many instances is recoverable. In some of this underground environment, conditions occurred occasionally which must have been favourable for the production of gemstone. The passage of immeasurable periods of time combined with these conditions to favour the formation of opal as well as other gemstones such as ribbonstone and quartz-type minerals.

When the total area of the Artesian Basin and the previous inland sea is taken into consideration, it becomes reasonably obvious that there must be vast deposits of precipitated minerals in original state of deposition still awaiting discovery by some fortunate fossicker. This is to say nothing of the great mineralisation of the volcanic and metamorphised areas of the continent which have only been superficially scratched over the past two centuries. Much still remains to be learnt of these deposits and the attention of the fossicker could well be turned in this direction.

Australia offers opportunities to the gemstone fossicker which are almost unequalled anywhere else in the world, and all that is required for success is an intelligent, commonsense approach, and a modicum of luck.

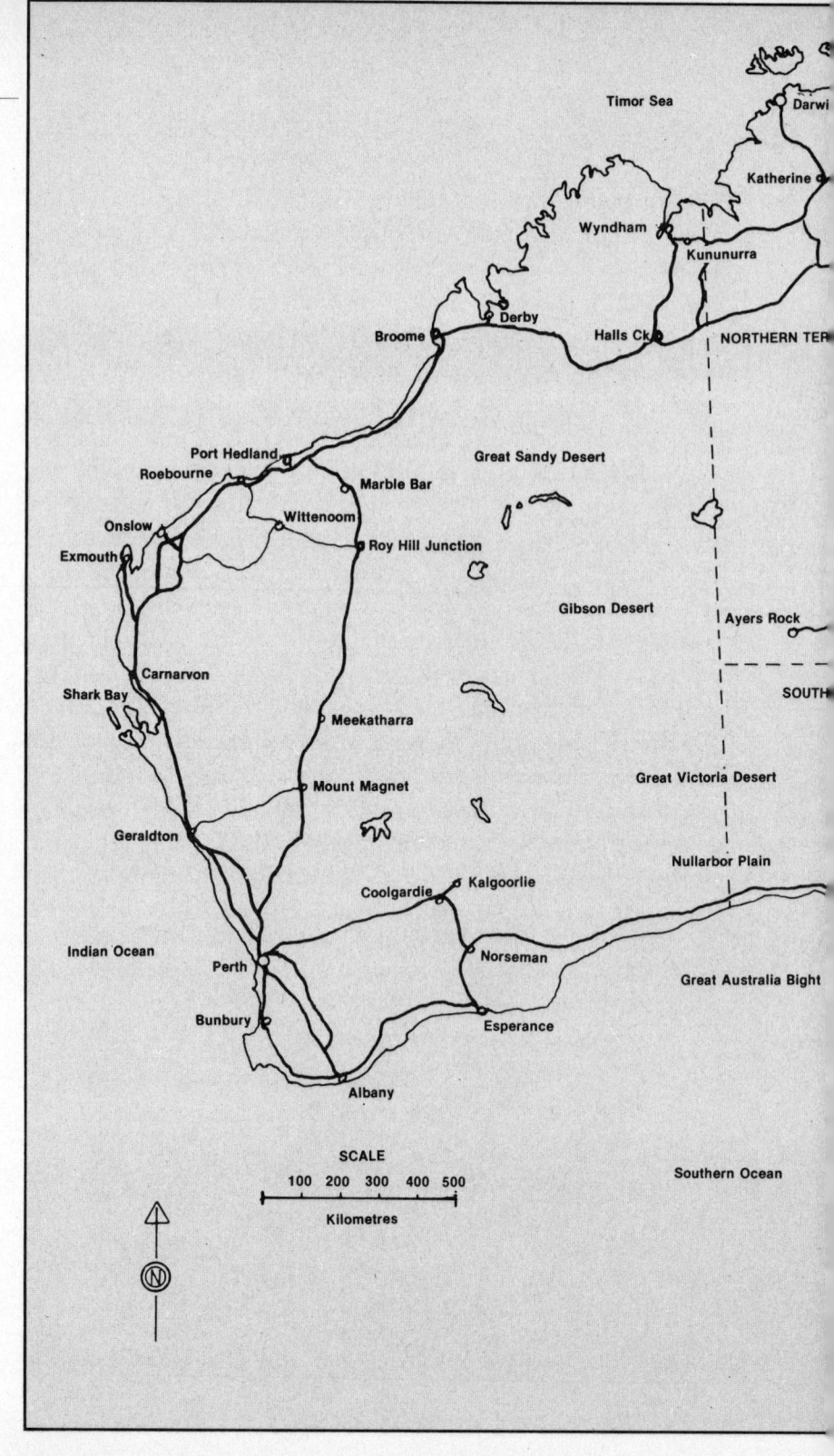

Principal gemstone areas of Australia, with access roads shown.

43

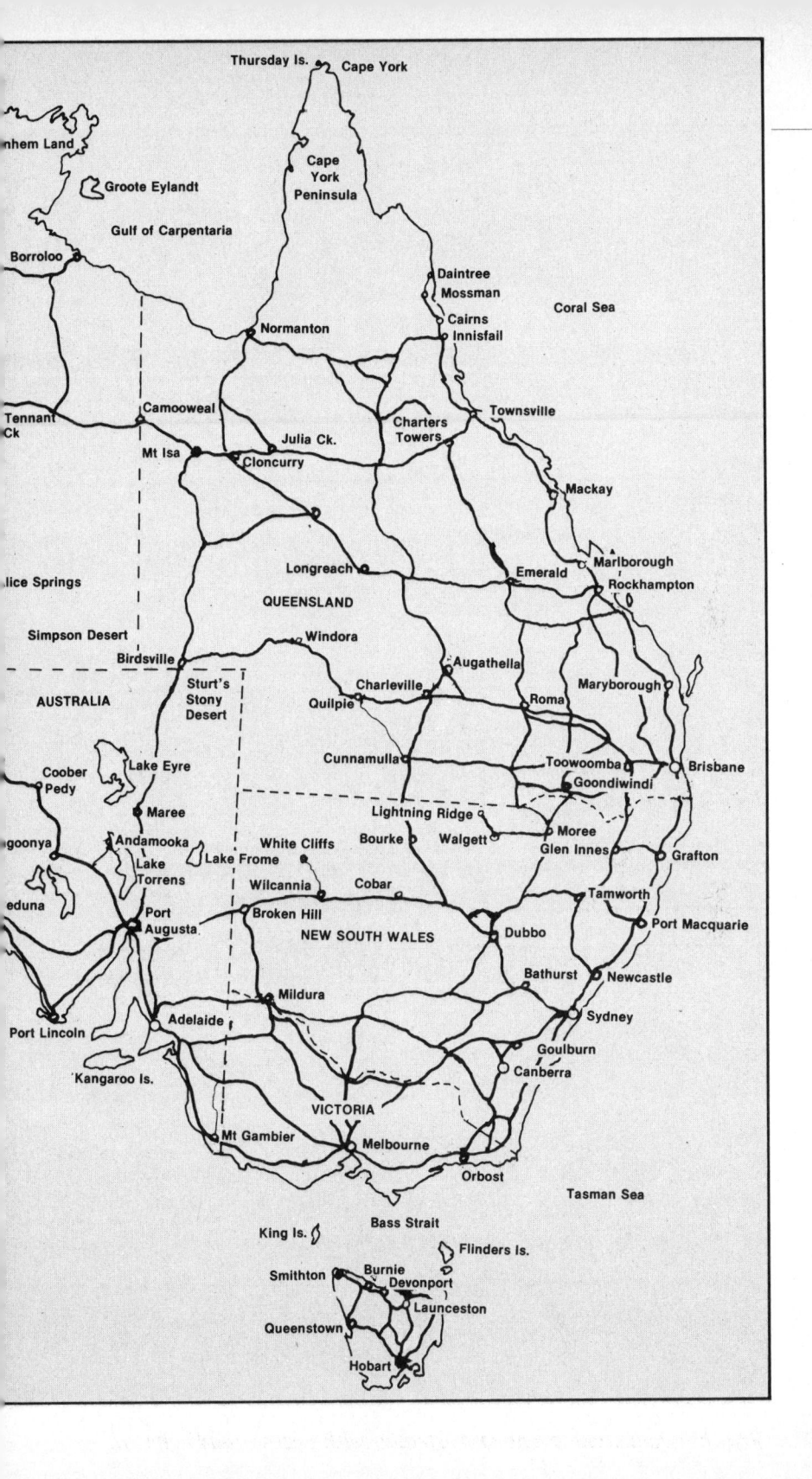

All states guide to gemstone localities

Actinolite

NSW Giant's Den Knob (near Bendemeer).
SA   Arkaroola, Olary Province.

Agate

NSW Baan Baa, Bald Hill, Baradine, Barraba, Bellata, Berridale, Boggabri, Bowning, Bugaldie, Cobham, Collarenebri, Cowriga, Cumborah, Doon Doon, Drake, Dubbo, Goorianawa, Grove Creek, Gunningbland, Gwydir River, Hunter River, Kangaroo Valley, Kiama, Lismore, Macintyre River, Monaro, Moree, Mount Wingen, Murrumburrah, Murwillumbah, Narrabri, Nundle, Oban, Quirindi, Singleton, Therabri, Tingha, Trunkey, Tweed River, Wee Waa, Wellington, White Rock (near Drake), Wollongong.
Qld Agate Creek (50 kilometres south of Forsayth), Calliope River, Cedar Creek, Chillagoe, Cloyna, Croydon, Etheridge, Gilberton, Herberton, Little River (48 kilometres east of Croydon), Longreach, Mitchell River (90 kilometres north of Chillagoe), Mount Hay, Murgon, Nanango, Nerang River, Percyville, Proston, Redcliffe, Williams, Windera (15 kilometres north of Murgon).

Vic.   Beechworth, Buchan district, Casterton, Castlemaine, Derrinal, Derrinallum, Eldorado, Gellibrand River (mouth of), Glenrowan, Mooloort, Moonlight Head, Murray River, Ovens River, Phillip Island (beaches at), Yarra River (upper reaches).

WA   Agate Hill, Antrim Plateau, Bamboo Springs, Halls Creek, llgarari, Moora, Mount Deception, Mount Frank, Mount Herbert, Spargoville, Wandagee.

Tas.   Cape Portland, Cornelian Bay, Ilfracombe, Lymington, Mangalore, Southport.

Alunite

NSW   Buledelah.

Amazonite

SA   Arkaroola, between Tumby Bay and Port Lincoln, Moonta.

WA   Payne's Find.

Amethyst

NSW   Boggabri, Broken Hill, Glen Elgin, Kingsgate, Mittagong, Native Dog Creek, Newstead, Oban, Oberon, Tingha, Torrington.

Qld   Anakie, Beechmont, Cloyna, Kingar Creek, Kuridala, Logan River, Lowmead, Rocky Creek, Sapphire, Stanthorpe, Tomahawk Creek, Williams, Windera.

Vic.   Beechworth, Bendigo, Berwick, Cardinia Creek, Castlemaine, Eldorado, Linton, Maldon, Mount William, Ovens River district, Whitfield.

SA   Ashburton River, Iron Knob (nearby area), Mount Augustus (21 kilometres southwest of), Mount De Courcey (8 kilometres southeast of), Murchison River, Sim Creek.

Tas.   Big Grassy Hill (Cape Barren Island), Blue Tier, Cape Portland, Derby, Emu River (south of Hampshire), Gladstone, Lefroy district, Moorina, Mount Cameron, Rossarden, South Mount Cameron.
NT   Harts Range, Plenty River area.

Andalusite

Qld   Mount Isa district.
NSW   Kempsey.
SA   Bimbowrie, Victor Harbour.
WA   Mungari, Nevoria.

Apatite

SA   Angaston, Bimbowrie (black variety), Mount Painter, Myponga, Tourmaline Hill.
NT   Jemkin's Old Camp (near Mud Tank, between Harts Range and Stuart Highway).
Tas.   Blue Tier, Cape Portland, Mathinna, Mount Bischoff.

Aquamarine

NSW   Emmaville, Inverell, Torrington.
Qld   Quart Pot Creek.

Aragonite

NSW   Bugaldie, Goorianawa, Kulnura, Manilla, Unanderra.
Vic.   Collingwood.
SA   Morgan.

Australite (tektite)

WA   Southeastern areas.
SA   Widespread.
Vic.   Northwestern areas.
Tas.   Northwestern areas.
NT   Widespread.

Axinite

Tas. Colebrook Mine (northeast of Dundas), Mount Ramsay, Parkes Hood.

Azurite

NSW Broken Hill, Cobar, Drake, Emmaville.
Qld Clermont, Cloncurry, Mount Isa, Peak Downs.
SA Arkaroola, Burra Burra, Moonta, Pernatty Lagoon.
WA Cue, Hammersley Gorge, Kalgoorlie, Norseman, Ravensthorpe, Southern Cross.
NT Daly River, Jervois Range, Rum Jungle, Tennant Creek.
Tas. Dundas, Hampshire, Heazelwood River, Mainwaring Inlet, Mount Lyell.

Beryl

NSW Black Range, Blatherarm Creek, Broken Hill, Bungonia, Cooma, Cow Flat, Crookwell, Elsmore, Emmaville, Frasers Creek, Glen Creek, Glen Eden, Glen Innes, Kangaroo Flat, Kiandra, Ophir, Paradise Creek, Scrubby Gully, Stanborough, Stannum, Surface Hill, Tingha, Torrington, Tumbarumba, Vegetable Creek, Wunglebung.
Qld Broadwater Creek, Decloath Creek, Galah Creek (Mount Isa), Hunts Creek, Kettle Swamp Creek, Lode Creek, Mica Creek, Quartpot Creek, Quartz Hill, Sugarloaf Creek.
Vic. Beechworth, Maldon, Pakenham.
WA Jilbadja, Melville, Poona, Spargoville, Wodgina, Yinnietharra.
NT Daly River, Disputed Mine (Harts Range), Jervois Range.
Tas. Bell Mount, Flinders Island, Great Republic Mine (Ben Lomond), Moina and along Saint Pauls River (opposite Brookstead).

SA   Williamstown.

Cairngorm

NSW   Gilgai, Glen Innes, Torrington.
Vic.   Maldon.

Carnelian

NSW   Bellata, Wee Waa, Werris Creek (east of town).
SA   Mount Gee, Radium Ridge.

Cassiterite

NSW   Tingha.
NT   Alice Springs area, Harts Range (black variety).
Qld   Stanthorpe.
Tas.   Coles Bay, Flinders Island, Goshen, Mount Cameron, Rossarden, Waratah.

Chalcedony

NSW   Bendemeer, Bingara, Boggabri, Carcoar, Coalcliff Beach, Doon Doon, Dubbo, Garie Beach, Gulgong, Gunnedah, Hunter River, Lue, Maitland, Narrabri, Newstead, Norah Head, Nundle (Munro's Creek and Bowling Alley Point at head of Peel River), Oban, Quirindi, Richmond River, Somerton, Swanville, Tooraweena, Tweed River, Tyalgum, Wee Waa, Wellington, Werris Creek, Wollongong.
Vic.   Beechworth district, Cape Otway, Casterton, Dookie, Gellibrand River (mouth of), Goulburn River, Heathcote, Mooloort, Moonlight Head, Moroka River, Murray River, Nowa Nowa, Ovens River, Phillip Island, Tatong, Yandoit, Yarra River.
Qld   Agate Creek, Anakie, Murong, Nerang River, Percyville.
Tas.   Cape Portland, Cornelian Bay, Flinders Island,

Goulds Country, Lake Sorrell, Lisle, Meredith Range, Mount Cameron.

Chiastolite

WA   Mungari, Nevoria.
SA   Bimbowrie, Mount Howden, Victor Harbour.
Qld   Mount Isa district.

Chrysoberyl

Vic.   Beechworth, Wooragee.
Tas.   Weld River (alexandrite variety).

Chrysocolla

NSW   Cobar.
Qld   Mount Isa area.
SA   Arkaroola, Burra Burra, Ethiudna Mine (on Plumbago Station).

Chrysoprase

NSW   Beaches between Coalcliff and Kiama.
Qld   Kilkivan area (40 kilometres west of Gympie), Marlborough Creek area (80 kilometres northwest of Rockhampton).
WA   Bulong district, Comet Vale, Wingellina.
SA   Mount Davies area.

Citrine

NSW   Emmaville, Gilgai, Glen Innes, Kingsgate, Murrumbidgee River, Oban, Red Range, Rockley, Tingha area (widespread), Torrington, Uralla.
Qld   Mount Isa area.
Vic.   Ararat, Beechworth, Eldorado, Maldon.
NT   Yambah Station.
Tas.   Goulds Country, Moorina, Mount Cameron.

Common opal

NSW	Barraba, Carcoar (Rocky Bridge Creek), Coonabarabran, Delungra–Warialda district, Folleys Creek, Hanging Rock, Munro's Creek (Nundle), Spring Creek, Tintenbar, Tooraweena.
Qld	Blackwater, Buderim Mountain, Cedar Creek, Chinchilla Station, Cooper Creek, Cooram, Cordalba, Mount Britten, Mount Sylvia, Mount Toussaint, Pelican Creek, Pine Mountain, Scotland Hills, Stanwell, Swanfels Valley.
Vic.	Bacchus Marsh, Baringhup, Bass River, Beechworth, Daylesford, Gelantipy, Gisborne,

A thirty-kilogram piece of green chrysoprase from Marlborough, Queensland.

	Grampian Mountains, Keilor, Kyneton, Malmsbury, Mirboo, Mornington Peninsula, Mount Stavely, Sassafras Creek, Sunbury, Wellington River, Williamstown, Woori Yallock.
SA	Angaston (between Cowell and Cleve), Williamstown, Wilpena Hill.
WA	Bulong, Grants Patch, Kennedy Range, Lake Rebecca, Lionel, Mundimindi, Norseman (north of), Ord Range, Paris Mining Centre, Poona, Widgiemooltha, Wittenoom area, Yarra Yarra Creek, Yerila, Yundamindra.

Crocoite

Tas.	Dundas Mine, Zeehan.

Diamond

NSW	Airly Mountain, Ashford, Ballina, Barraba, Bathurst, Bendemeer, Bingara, Boggabri, Calula Creek, Copes Creek, Copeton, Crookwell, Cudgegong River (near Gulgong), Delungra, Dubbo, Emmaville, Euriowie, Grabben Gullen, Hill End, Howell, Inverell, Macquarie River (near Wellington), Mittagong, Mount Werong, Mudgee, Nandewar Range area, Narrabri, Native Dog Creek, Oberon, Pine Ridge, Pyramul Creek, Shoalhaven River, Turon River and Reedy Creek (near Bathurst), Upper Tarlo, Uralla, Wellington, Wheeo, Wingecarribee.
SA	Algebuckina, Echunga goldfields.
Qld	Anakie, Glenalva, Herberton, Rubyvale, Sapphire, Stanthorpe, Tomahawk Creek, Willows.
WA	Kimberley region, Lennard River (northern Canning Basin), Nullagine.
Vic.	Beechworth, Castlemaine, Chiltern, Eldorado, Mansfield, Wooragee.
Tas.	Savage River.

Emerald

NSW	Emmaville, Frasers Creek, Guyra, Kiandra, Oban, Paradise Creek, Tingha, Torrington, Tumbarumba.
Vic.	Daylesford, Donnellys Creek, Upper Yarra.
WA	Calverts White Quartz Hill, Cue, McPhees Patch, Melville, Pilbara goldfields at Wogina, Poona and Warda Warra (in Murchison goldfields).
Tas.	Thureau's Deep Lead (near Saint Helens).

Epidote

SA	Olary Province.
WA	Hatters Hill.
Tas.	Deloraine, Dundas, Emu River, Mainwaring Inlet, Mount Bischoff, Table Cape.
NT	Harts Range.

Feldspar

Vic.	Castlemaine.
NT	Harts Range.

Fluorite (fluorspar)

NSW	Bowling Alley Creek, Broken Hill, Glen Innes, Horsearm Creek (near Attunga), Mount Eltie, Mount Robe, Sheep Station Creek (Nundle area).
Qld	Almaden (north of), Bullock Creek area, Dargalong, Emuford, Herberton, Mary Kathleen (8 kilometres east of), McCords (north of Fisherton), Mount Garnet, Mungana.
SA	Moonta, Mutooroo, Pernatty Lagoon (clear variety), Plumbago (purple variety).
Tas.	Babel Island, Great Republic Mine (Ben Lomond), Hampshire, Lottah, Mount Bischoff, Mount Ramsay, Rosebery, Zeehan.

Garnet

NSW Abercrombie, Adjinbilly, Albury, Attunga, Barraba, Bathurst, Beardy Waters, Bingera, Boggabri, Bowling Alley Point, Broken Hill, Bullock Mountain, Carcoar, Conona, Copeton, Crookwell, Fish River, Glen Innes, Grafton, Gulgong, Gundagai, Guyra, Hardwicke, Hartley, Inverell, Kingsgate, Macquarie River, Mittagong, Moama, Mount Tennyson, Mudgee, Murrumburrah, Native Dog Creek, Nundle, Oban, Oberon, Pambula, Pond's Creek, Poolmacca, Reddestone Creek, Red Range, Ruby Hill, Sidmouth Valley, Silverton, Tallong, Tamworth, Thackaringa, Trunkey Creek, Tumbarumba, Uralla, Wallerawang, Warialda, Washpool Creek, Wee Jasper, Whipstick, Wingecarribee River, Yass, Yetholme.

NT Alice Springs (northeast of), Harts Range, Jervois, MacDonnell Ranges (east of).

Vic. Ararat, Ballarat, Barnawatha, Beechworth, Berwick, Blackwood, Castlemaine, Chiltern, Eldorado, Glendinning, Harrow, Lilydale, Longford, Maldon, Ovens, Point Leo, Reedy Creek, Stawell, Tallangatta, Toombullup, Tubbarubba, Wodonga, Woolshed Creek, Yanakie.

SA Arkaroola, Ethiudna Mine (on Plumbago Station), Kanmantoo, Millendella, Mount Painter, Nairne, Tourmaline Hill.

WA Cooglegong, Moolyella, Nornalup, Northampton, Rothsay, Yabberup, Yinnietharra.

Tas. Bellmount, Comstock, Cygnet, Emu River, Grassy (on King Island), Hampshire, Heazelwood River, Hudson and Lewis Rivers, Maynes Tin Mine (south of Mount Heemskirk), Moina, Mount

	Claude, Mount Kerford (Cape Barren Island), Mount Ramsay, Mount Stewart, Sea Elephant Point, Stony Ford (2 kilometres from Georges Bay), Trial Harbour, White River.
Qld	Blackbutt (Googa Creek), Booreeco Creek (near Barmundoo), Brigooda (19 kilometres west of Proston), Broken River, Bunerba, Cattle Creek (Corella River), Cloncurry (south of), Coondarra Creek (headwaters and streams which flow into Lawson's Broad), Diglum Creek, Dry River, Emuford, Eungella, Glassford Creek (near Many Peaks), Glenbar, Innisfail, Jordan Creek, Jordan Mining Field (39 kilometres west-south-west of Innisfail), Lowood (45 kilometres west of Brisbane, nearby areas), Maronan, Mitchell River, Morinish, Mount Garnet, Mount Grim, Mount Isa, Mount Surprise, Mount Tarampa, Munna, Murphys Creek, Newcastle Range, Opera Creek, Percy River, Pinevale, Proston, Quartz Hill (near Dagworth), Redcap Creek, Silver Valley, Spring Creek, Tarampa, Upper Cattle Creek area, Westwood.

Goshenite

See Beryl.

Heliodor

See Beryl.

Hematite

NSW	Carcoar.
SA	Angaston (near Nairne), Arkaroola Bore area, Moonta–Wallaroo (Yorke Peninsula).
WA	Hamersley Iron Province, Yampi Sound.

Horn (buffalo)

NT   Widespread.

Hyalite

NSW   Mullumbimby.
Qld   Cooram, Minerva Creek, Square Top Mountain.
Vic.   Baringhup, Gisborne, Kyneton, Malmsbury.
SA   Cowell.

Iolite

NT   Harts Range.
SA   Mount Pitt.

Jade

See **Nephrite jade** *and* **Pilbara jade**.

Jasper

NSW   Baan Baa, Baradine, Barraba, Bellata, Bingera, Boggabri, Bugaldie, Collarenebri, Cumborah, Hanging Rock, Inverell, Kangaroo Valley, Quirindi (brecciated jasper), Shoalhaven River, Tamworth, Taree, Warialda.
Vic.   Beechworth, Cape Otway, Castlemaine, Dookie, Goulburn River, Heathcote, Murray River, Nowa Nowa, Tatong.
WA   Marble Bar, Weld Range.
SA   Arkaroola area.

Jaspilite

WA   Hamersley Iron Province.
SA   Iron Knob, Wilpena Hill.

Jet

NSW   Guy Fawkes (north of).

Kyanite

NSW   Thackeringa.
NT   Harts Range.

Labradorite

NSW   Hogarth Range, Mullumbimby area.
Qld   Tambo Road (out from Springsure).

Lazulite

SA   Montaro.

Malachite

NSW   Blayney, Cobar, Girilambone, Mount Hope.
Qld   Mount Isa area.
SA   Arkaroola, Burra Burra, Moonta Mines (Yorke Peninsula), Mount Painter, O'Donahues Castle (on Balcanoona Station), Pernatty Lagoon (north of Gawler Range).
Tas.   Badger Head, Cascade River, Franford, Heazlewood River, Mainwaring Inlet, Mount Lyell, Scamander River, Zeehan.

Marcasite

Vic.   Morwell.
WA   Northampton.
Tas.   Cape Barren Island.

Mookaite

WA   Gascoyne River area.

Moonstone

NSW   Bolivia (south of Tenterfield), Hutchinsons Lode (4 kilometres southeast of Tingha).
Qld   Bell, Greenvale, Minerva Creek, Porters Gap, Springsure.
Vic.   Castlemaine.

Morganite

See Beryl.

Morion

NSW   Tingha, Torrington.

Nephrite jade

NSW   Dungowan (24 kilometres east of Tamworth), Lucknow.
SA   Eyre Peninsula (21 kilometres northeast of Cowell).

Obsidian

NSW   Boggabri.

Onyx

See Agate *and* Chalcedony.

Opalised wood

NSW   Barraba, Berigal Creek area, Bugaldie, Coonabarabran, Edgeroi, Goorianawa, Narrabri, Warialda.

Opalite

NSW   Manilla, Port Macquarie.
Vic.   Beechworth, Cardinia Creek.
WA   Carnarvon, Kimberley region.

'Peanut' jasper

Qld   Agate Creek.

Peridot

NSW   Gum Flat Road (6.5 kilometres from Inverell).
Qld   Atherton Tableland (between Cheviot Hills and Lyndhurst Homesteads, 255 kilometres west of Townsville), Main Range (vicinity of

	Toowoomba), Spring Bluff (10 kilometres north-north-east of Toowoomba, nearby area).
Vic.	Bendigo, Camperdown (bombs), Daylesford, Maldon, Mount Lookout.
SA	Mount Davies (brown), Mount Gambier (in volcanic bombs).
Tas.	Deloraine, Doctors Rocks, Don Heads, East Arm (Tamar River), Emu River, Hampshire, Mount Wellington, Scottsdale, Sidling, Upper Forth River, Waratah – Wilmot River area.

Petalite

WA	Londonderry (near Coolgardie).

Petrified wood

NSW	Baan Baa, Baradine, Barraba, Bellata, Berigal Creek area, Bingera, Boggabri, Bugaldie, Collarenebri, Copeton, Cumborah, Dubbo, Edgeroi, Goonoo Goonoo, Goorianawa, Gunnedah, Lake Macquarie (western shores), Milparinka, Nandewar Range area, Narrabri, Tilbuster, Walcha, Warialda, Wollongong (north of).
Tas.	Cameron, Derby, Hobart region, Lake Sorrell, Latrobe, Launceston, Lisle, Longford.

Pilbara jade (massive green chlorite)

WA	Five Mile Well (northeast of Roebourne), Marble Bar, Nullagine (25 kilometres north of Nullagine).

Prase

NSW	Nundle.
WA	Spargoville.

Precious opal

NSW	Angledool, Grawin, Lightning Ridge, Tintenbar, White Cliffs.

SA   Andamooka, Coober Pedy, Mintabie.
Qld   Black Gate, Bulls Creek, Coonavilla, Duck Creek, Elbow, Emu Creek, Fiery Cross, Grey Range, Jundah, Koroit, Kynuna, Opalton, Quartpot Creek, Quilpie, Sheep Station Creek, Springsure, Toompine, Valdare, Winton, Yowah.
WA   Coolgardie (4.8 kilometres north-north-east), Cowarna Station (13 kilometres east).

Prehnite

NSW   Gunnedah, Prospect, Tambar Springs.
WA   Comet Vale, Coolgardie, East Kimberley region, Mount Palmer.

Pseudomalachite

SA   Spring Creek Copper Mine (Melrose).

Psilomelane

Vic.   Buchan district.
WA   Horseshoe area (130 kilometres north of Meekatharra).

Pyrite

NSW   Broken Hill.
SA   Arkaroola area, Brukunga, Moonta, Mutooroo (south of Cockburn).
Tas.   Cape Barren Island, Cox Bight, Magnet Mine, Mount Lyell, Scamander River.

Rhodochrosite

NSW   Broken Hill.
Tas.   Dundas, Magnet, Mount Reed, Rosebery, Zeehan.

Rhodonite

NSW   Bendemeer, Broken Hill, Copeton, Danglemah, Kempsey, Moonbi Range (especially Hall's Creek),

	Niangla, Niangla Road–Duncan's Creek area, Nundle, Port Macquarie, Tamworth, Tingha, Walcha.
Qld	Warwick.
WA	Hamersley Range, Roebourne.
Tas.	Anderson's Creek (west of Beaconsfield).

Ribbonstone

WA	Hamersley Range, Roebourne.
Qld	Barkly Tableland, Camooweal.
NT	Anthony's Lagoon, Tennant Creek area.

Rock crystal

NSW	Abercrombie River, Bendemeer, Bingara, Drake, Emmaville, Glen Innes, Guyra, Hanging Rock (near Nundle), Inverell, Kingsgate, Newstead, Oban, Peel River, Torrington, Tuena.
Vic.	Castlemaine, Maldon.
Qld	Fischerton.

Rose quartz

NSW	Hall's Creek (near Moonbi).
Qld	Fischerton.
Tas.	Beaconsfield, Blue Tier, Lefroy, Moorina.

Ruby

NSW	Crookwell, Gulgong, Hill End, Horse Gully, Inverell, McGlanglin Creek (near Nimmitabel), Macquarie River (near Wellington), Mudgee, Native Dog Creek, Oberon, Wandsworth area (23 kilometres northeast of Guyra).
Vic.	Beechworth, Beenak, Berwick, Castlemaine, Daylesford, Glendinning, Foster, Mount Eliza, Pakenham, Tanjil, Toombullup, Toora, Tubbarubba.
Tas.	Boat Harbour, Weld River.

SA   Mount Pitt.
NT   Harts Range.

Rutilated quartz

NSW   Glen Innes, Guyra, Inverell, Tingha.

Sapphire

NSW   Abercrombie River, Bald Nob, Barraba, Beardy Waters, Ben Lomond, Berrima, Bingara, Bullock Mountain, Copes Creek, Copeton, Crookwell, Cudgegong River, Duckmaloi River, Dundee, Ebor, Elsmore, Frasers Creek, Glencoe, Glen Elgin, Glen Innes, Grabben Gullen, Green Hills, Guyra, Gwydir River, Hanging Rock, Hill End, Horse Gully, Inverell, Mann River, Mary Ann Creek, Mittagong, Mole Tableland, Mount Werong, Mudgee, Namoi River, Native Dog Creek, Newstead, Niangla, Niangla Road – Duncan's Creek area, Nullamanna, Nundle (Peel River gravel beds), Oban, Oberon, Porters Retreat, Reddestone Creek, Sapphire, Tingha, Tumbarumba, Warialda, Wee Jasper.

Qld   Almaden, Anakie, Beatrice River, Blackbutt, Bunerba, Burrandowan Station, Central Creek, Chuddleigh Park, Drummond Range, Gilbert River, Glendarriwell, Goomburra, Henrietta Creek, Herberton, Hut Creek, Jordan Creek, Kettle Creek, Medway Creek, Mount Hoy, Murphys Creek, Nananga (Seven Mile Diggings), New Rush, Retreat Creek, Rubyvale, Russell goldfields, Sapphire, Serpentine Creek, Spring Creek, Stanthorpe, Tomahawk Creek, Willows, Withersfield.

Vic.   Ararat, Ballarat, Beechworth, Berwick, Blackwood, Cardinia Creek, Castlemaine,

	Chiltern, Colac, Daylesford, Dereel, Donnellys Creek, Foster, Gembrook, Glendinning, Kangaroo Crossing (near Eldorado), Leongatha, Mount Eliza, Mount Lookout, Pakenham, Point Leo, Tolmie, Toombullup, Toora, Tubbarubba, Upper Yarra.
SA	Mount Crawford, Mount Pitt.
WA	Byro Station (northwest of), Dangin, Jacobs Well, Mount Broome, Richenda River.
Tas.	Blythe River, Boat Harbour, Branxholm, Coles Bay, Derby, Gladstone, Lisle, Lottah, Main Creek, Moorinna, Mount Cameron, Mount Stronach, Sisters Creek, Stanley River tin field, Table Cape, Thomas Plains, Weld River.
NT	Harts Range, Plenty River.

Serpentine

NSW	Hanging Rock.
Tas.	Beaconsfield.

Smoky quartz

NSW	Elsmore (near Tingha), Gilgai, Kingsgate, Oban, Torrington.
Vic.	Beechworth, Lake Boga, Strathbogie Ranges, Tarrangower, Upper Yarra.
SA	Moonta.
NT	Disputed Mine (Harts Range).
Tas.	Cox Bight, Derby, Dundas, Flinders Island, Savage River, Stanley River.

Sphene (titanite)

NSW	Broken Hill.
SA	Radium Creek.
NT	Harts Range.
Tas.	Cygnet, Heazlewood River, Mount Ramsay, Parsons Hood.

Spinel

NSW	Bathurst, Bingara, Copeton, Glen Innes, Inverell, Mudgee, Native Dog Creek, Oberon, Tingha, Tumbarumba, Uralla.
Vic.	Beechworth, Castlemaine.
Tas.	Branxholm, Cygnet, Derby, Gladstone, King Island, Moorina, Mount Bischoff, Rossarden, Thomas Plains.

Spodumene

WA	Kalgoorlie (south of), McPhee Range.

Steatite

Vic.	Kevington.
NSW	Botolbar, Bucharoo, Harilah, Mudgee district.
Tas.	Gawler.

Stichtite

Tas.	Dundas, Serpentine Hill (northwest of Birch Inlet, off Macquarie Harbour).

Tektite

See **Australite**.

Thunder egg

NSW	Baan Baa, Boggabri, Doon Doon, Murwillumbah, Perch Creek.
Qld	Agate Creek, Beechmont, Mount Hay, Tamborine Mountain.
Vic.	Buchan district, Snowy River (lower reaches).

Tigereye

WA	Hamersley, Mount Newman, Mount Tom Price, Wittenoom.

A cut thunder egg showing an infilling of agate and quartz crystals.

Topaz

NSW   Armidale, Backwater, Bald Nob, Barraba, Bathurst, Beardy Waters, Bendemeer, Bingara, Boggabri, Boonoo Boonoo, Bullock Mountain, Collarenebri, Copeton, Crookwell, Cudgegong River, Cumborah, Dundee, Emmaville, Frasers Creek, Glen Creek, Glen Innes, Gulgong, Gundagai, Inverell, Kingsgate, Kookabookra, Lightning Ridge, Mittagong, Mudgee, Oban, Reddestone, Shoalhaven River, Stannum, Tent Hill, The Gulf, Tingha, Torrington, Tungsten, Uralla.

Qld   Almaden, Anakie, Bald Creek, Ballendeen, Blackbutt, Broadwater Creek, Chillagoe, Coolgarra, Dry River (Newellton), Emuford,

	Etheridge, Gilberton, Henrietta Creek, Herberton, Hunts Creek, Innot Hot Springs, Jordan Mining Field, Kettle Swamp Creek, Lappa, Lode Creek, Mount Garnet, Mount Surprise, Nettle Creek, O'Brian's Creek, Pikedale, Quartpot Creek, Quartz Hill, Rocky Creek, Seven Mile Diggings (11 kilometres southeast of Nanango), Silver Valley, Stanthorpe, Sugarloaf Creek, Swipers Gully, Thulimbah, Wyberba.
Vic.	Ararat, Bacchus March, Beechworth, Beenak, Bunyip River (in gravels), Carisbrook, Castlemaine, Donnellys Creek, Dundly, Dunolly, Eldorado, Foster, Gembrook, Glendinning, Lal Lal, Maldon, Merino, Omeo, Pakenham, Stawell, Talbot, Tarrengower, Toora, Waratah, Yackendandah, Yarra River (upper reaches).
SA	Barossa Ranges, Kangaroo Island.
WA	Dangaranga, Londonderry (near Coolgardie), Melville, Poona, Wodgina.
Tas.	Beaconsfield, Bell Mount, Branxholm Creek, Brown Plains, Coles Bay, Derby, Dorset Flats, Gipps Creek, Gladstone, Killicrankie Bay (Flinders Island), Lefroy, Long Plains, Mathinna, Moina, Moorina, Mount Bischoff, Mount Cameron, Rossarden, Saint Pauls Creek, Stanley River, Thomas Plains, Weldborough, Weld River.

Tourmaline

NSW	Barraba, Bendemeer, Bingara, Boggabri, Bundarra, Glen Innes, Hall's Creek, Moonbi, Native Dog Creek, Oberon, Sidmouth, Silverton, Torrington.
Qld	Anakie, Boondoomlia Station (west of Proston), Glenalva, Henrietta Creek, Jordan Creek, Mount Isa, Sapphire, Tomahawk Creek, Watsonville, Willows.
Vic.	Beechworth, Cassilis, Castlemaine, Dandenong,

ALL STATES GUIDE TO GEMSTONE LOCALITIES

	Maldon, Mount Alexander, Saint Arnaud, Strathbogie Ranges, Upper Yarra, Waratah, Wilsons Promontory.
SA	Kangaroo Island, Moonta–Wallaroo, Mount Painter, Mount Pitt, Yorke Peninsula, Tourmaline Hill.
WA	Catlin Creek, Dalgaranga, Mount Hunt, Spargoville, Yinnetharra.
Tas.	Flinders Island, Lake Lea, Mount Bischoff, Mount Heemskirk, Mount Lyell, Mount Montgomery, Mount Ramsay, Stanley River.

Tourmaline in quartz

NSW	Emmaville, Gilgai, Kingsgate, Oban, Red Range, Rockley, Tingha, Torrington.
Qld	Mount Isa area.
NT	Harts Range (schorl).

Turquoise

NSW	Batemans Bay (south of), Bodalla, Grants Lookout (5 kilometres north-north-east of Narooma), Lake Mummuga, Murwillumbah.
Qld	Emu Park.
Vic.	Edi, Myrrhee, Whitfield.
SA	Mount Painter.
Tas.	Beaconsfield, Den Ranges (east Tamar), Lefroy – Back Creek area, Waddamana.
NT	Established mine 424 kilometres northeast of Alice Springs.

Variscite

Qld	Dayboro.

Zebra stone

WA	Kimberley region, Kunnurra region (inaccessible), Ord Range (near dam).

Zircon

NSW   Ann River, Apple Tree Gully, Armidale, Bald Nob (near Glen Innes), Barraba, Ben Lomond, Berrima, Bingara, Boggabri, Broken Hill, Copes Creek, Copeton, Crookwell, Cudgegong River, Duckmaloi Creek, Elsmore, Frasers Creek, Hanging Rock, Inverell, Mann River, Mittagong, Mudgee, Native Dog Creek, Niangla, Nundle, Oban, Oberon, Paradise Creek, Sapphire, Stoney Creek, Tilbuster, Tingha, Uralla, Wee Jasper, Wheeo, Wingecarribee River.

Qld   Agate Creek, Anakie, Beallah, Blackbutt, Broken River, Bunerba, Eungella, Fischerton, Goomburra, Greenhills (near Georgetown), Hazeldean, Henrietta Creek, Herberton, Herberton Creek, Jordan Creek, Kettle Swamp Creek, Murphys Creek, Pinevale, Rocky Creek, Rubyvale, Stanthorpe.

Vic.   Aberfeldy, Ballarat, Beechworth, Bendigo, Blackwood, Castlemaine, Chiltern, Colac, Daylesford, Donnellys Creek, Foster, Gembrook, Lal Lal, Lilydale, Leongatha, Merino, Omeo, Point Leo, Talbot, Tolmie, Toombullup, Toora, Tubbarubba, Wooragee.

SA   Kangaroo Island, Mount Painter.

WA   Greenbushes.

Tas.   Arthur River, Beaconsfield, Blythe River, Boat Harbour–Sisters Creek, Circular Head, Derby, Flinders Island, Gladstone, Long Island, Meredith, Moorina, Penguin, Rossarden, Ruby Flat, Thomas Plains, Trial Harbour, Weld River.

NT   Harts Range, Mud Tank, Strangways Range.

Zoisite

WA   Roebourne.

New South Wales

NEW SOUTH WALES has some of the best gem-fossicking spots in Australia within its borders, and it is not beyond the ability of individuals and clubs to visit these reasonably often.

On 21 August 1992 a new mining Act came into force. This Act has removed the necessity for fossicking licences in New South Wales. It allows fossicking to be carried out anywhere in the state provided no other Act or law prevents the carrying out of that activity on the land involved. Also, where private land is concerned fossicking may be carried out with the consent of the landholder. Furthermore, fossicking will be allowed within an area subject to an authority, a mineral claim or an opal-prospecting licence with the consent of the holder of the authority, claim or licence. It will be the responsibility of the intending fossicker to check whether the area he or she wishes to explore is available for fossicking. The Department of Minerals and Resources has a very comprehensive *List of Mineral Publications* dated 1995 which is available from Information and Customer Services. This list details many publications which are available, most for a fee. The principal item is *Gemstones* (1980, 120 pages, for $2). This contains a guide to gemstones in New South Wales. Major sections cover opal, diamond and sapphire; with information on beryl, silica gemstones, garnet, topaz, turquoise, zircon, spinel, sphene (titanite), rhodonite and nephrite (jade) also included. In addition there is a publication entitled *Location of*

A FOSSICKER'S GUIDE TO GEMSTONES

Fossicking Areas in New South Wales. This is available for free and comprises a series of sheets, in all over thirty pages of hand-drawn mud maps showing the permitted fossicking areas in the state in considerable detail, with access, topographical features and an abundance of other details included. A series of very useful pamphlets, *Minefact* by name, individually numbered, is also on free issue, each sheet highlighting many mineral-related facts. The Department's *Information Sheet No.102*, dated October 1992, gives a summary of the new mining Act.

The best known gem areas in New South Wales are

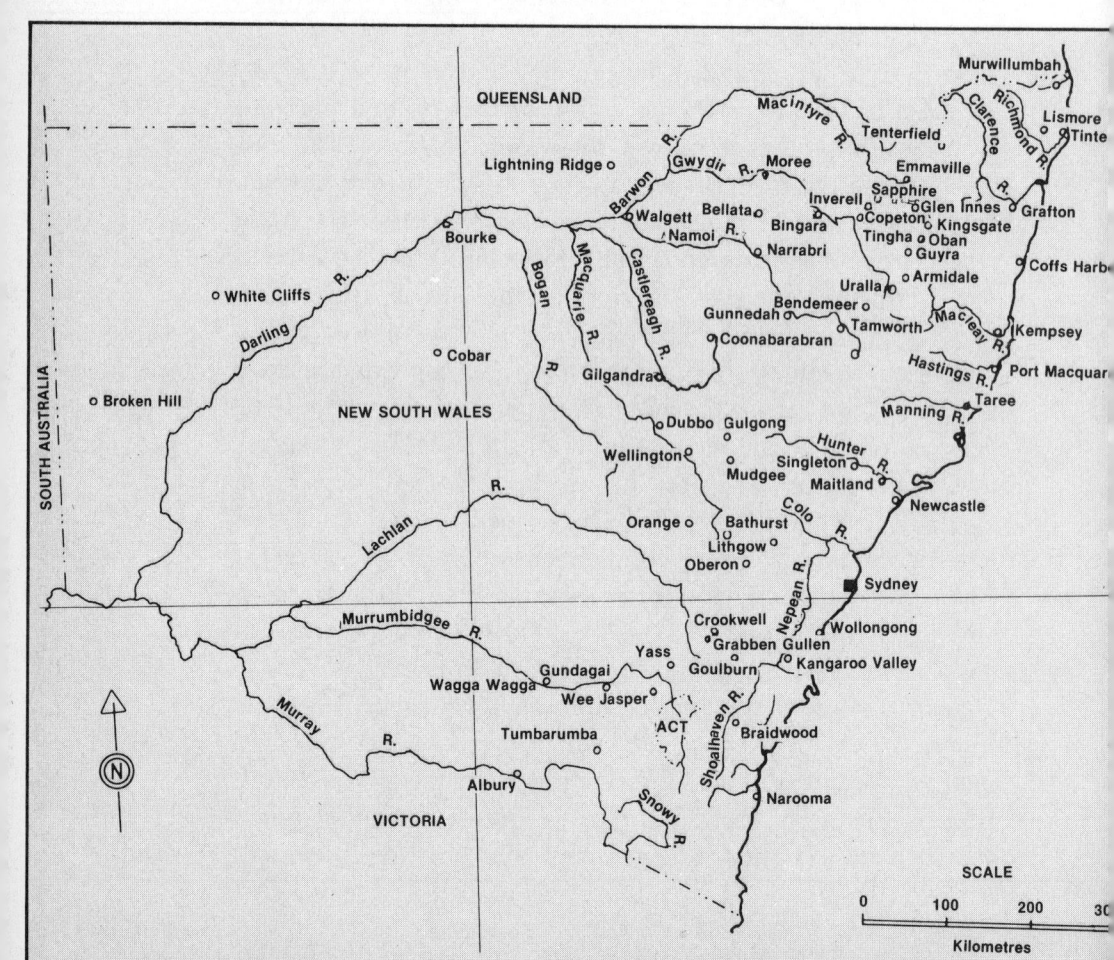

Principal rivers and gemstone-fossicking areas in New South Wales.

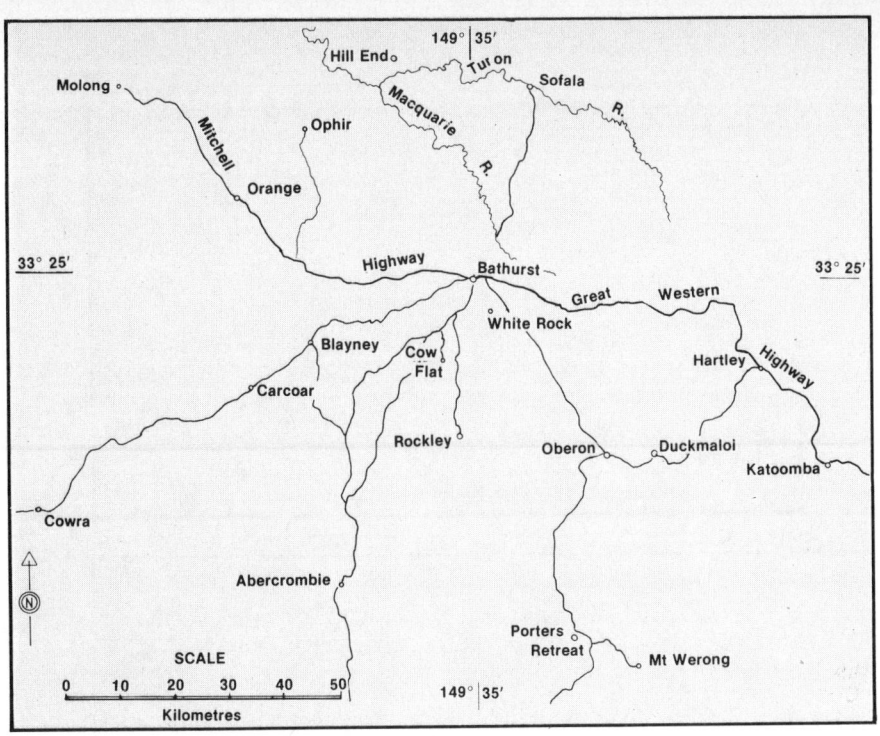

Gemstone-fossicking areas centred on Bathurst, central-western New South Wales.

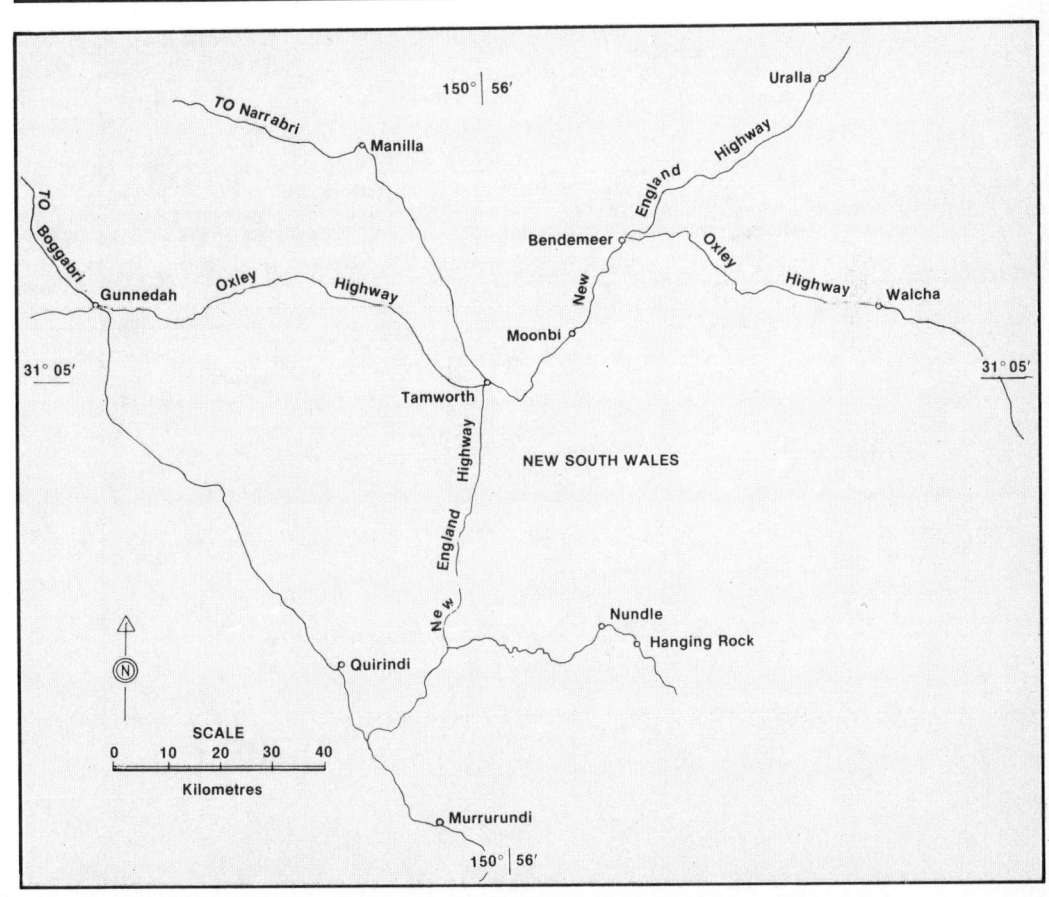

Gemstone-fossicking areas around Tamworth, with connecting roads shown.

Gemstone-fossicking areas centred on Goulburn, New South Wales, with main roads shown.

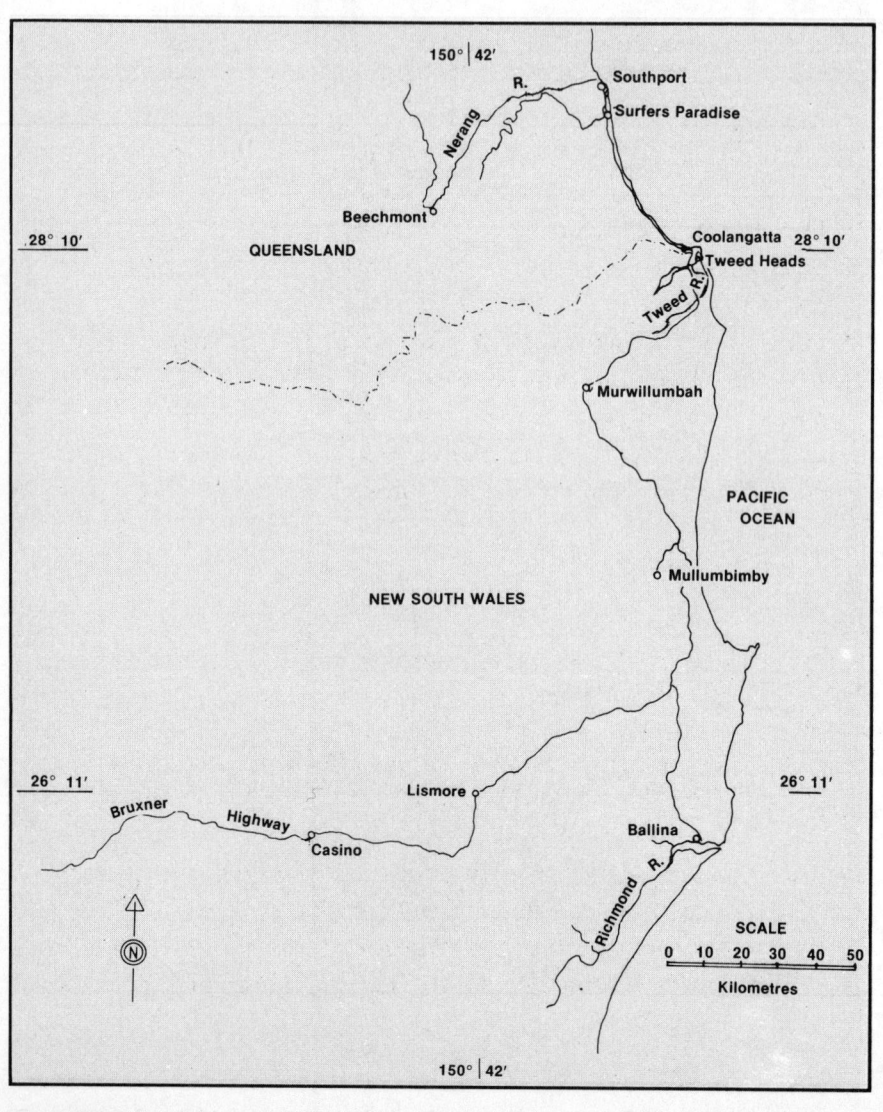

Gemstone-fossicking centres of the far north coast of New South Wales, with connecting roads shown.

Lightning Ridge for black opal, and the great New England District, where a large variety of gems are found, with sapphire holding pride of place.

Other gems which occur in New South Wales include agate, amethyst, beryl, feldspar, fluorspar, chalcedony, common opal, emerald, diamond, cairngorm, garnet, rhodonite, rock crystal, tourmaline, topaz, spinel and zircon.

The Kangaroo Valley in southern New South Wales offers a lot of interesting specimens for the fossicker, and because it is easily accessible is valued for family and club outings.

The Hanging Rock/Nundle area is another popular fossicking locality.

The Mittagong/Grabben Gullen district is particularly dear to the hearts of sapphire and diamond fanciers.

The Bathurst/Hill End/Oberon area provides endless interest for fossickers, with a wide variety of gems to be found.

The Northern Rivers district gives scope for the search of agate, chalcedony and thunder eggs.

The state's southern and central coast beaches are ideal for collecting waterworn agate and petrified wood, and the Port Macquarie area is a source of rhodonite.

New England district

ACCESS

Road Access by road is via the Gwydir Highway.
Rail Trains run to Glenn Innes.

ACCOMMODATION

Camping and caravan facilities are available throughout the Armidale/Inverell/Glen Innes region, as are top quality hotels and motels.

Many fossickers agree that the areas surrounding Inverell and Glen Innes in the New England district of New South Wales are the best prospecting locations in Australia. The gem fields are centred around Inverell, 195 kilometres west of Grafton, 35 kilometres west of Glen Innes, and 128 kilometres north of Armidale. Sapphire, garnet, topaz, zircon, spinel, beryl, aquamarine, emerald, tourmaline, ruby, amethyst, peridot, diamond, jasper, agate, rutilated quartz, and clear and smoky quartz crystals have all been found in this area.

The gems are located in river gravels, but because of the worldwide demand for sapphires, some commercial mining interests have closed off much of the land to the fossicker. These organisations are mining for sapphires with heavy plant and equipment, and their leases cover wide areas.

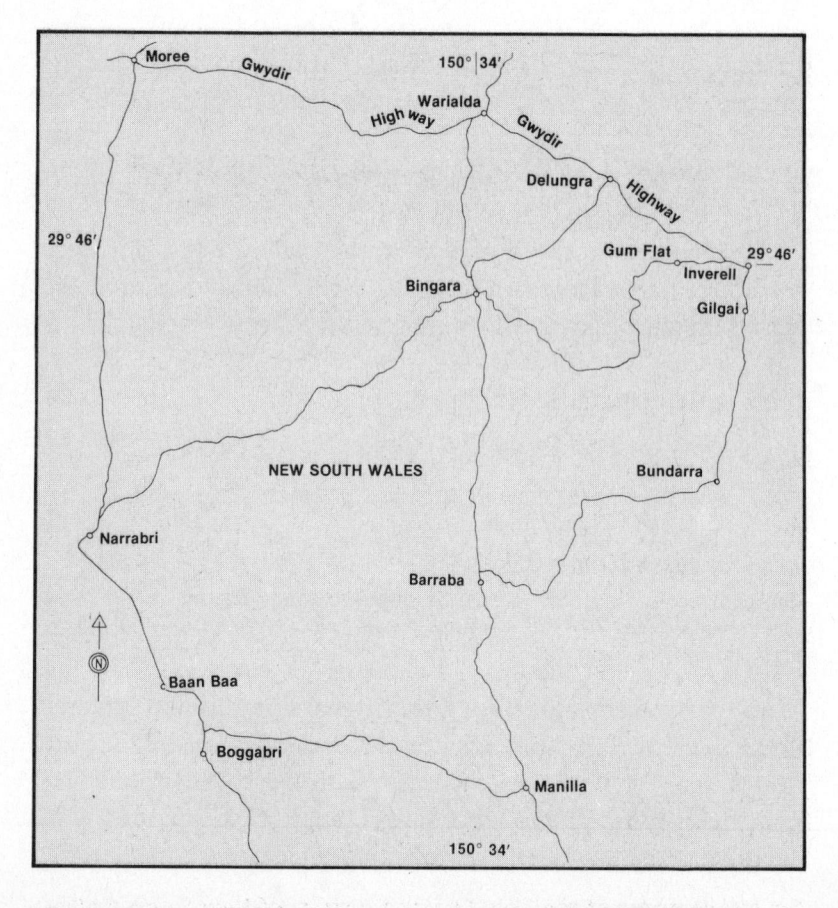

Gemstone areas of the western New England tableland, New South Wales, with main access roads shown. (Not to scale.)

However, there are still some areas set aside for the amateur to explore. In addition, sapphires and other gems may be searched for on privately owned farming property—with the landholder's permission of course.

The best times to visit the New England District are during the late spring, summer and autumn, as the winter weather is not too agreeable for fossickers, with rain, wind, and possibly snow occurring.

For on-the-spot advice on fossicking areas it is recommended that intending visitors contact the Bingara Shire Council in Maitland Street, Bingara.

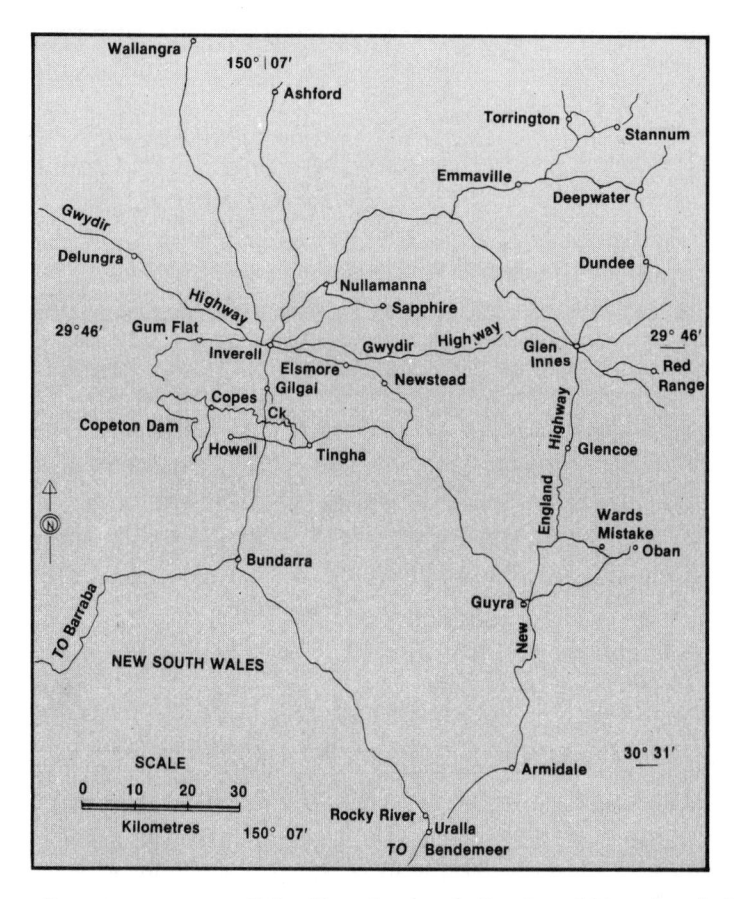

Gemstone areas of the New England district of New South Wales, with access roads shown.

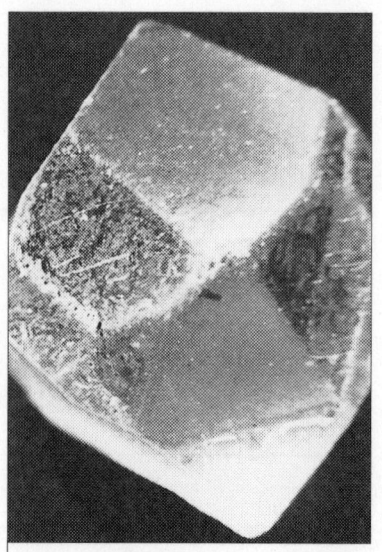

A well-shaped topaz crystal. This one has not been transported by a river or creek, as the sharper crystal edges bear witness.

A group of sapphire crystals showing the typical shapes to be found: some are large, distorted and less transparent, others are small but broken and clear or particolour. All could cut gems.

Always remember to obtain permission, preferably in writing, before entering private land for fossicking purposes.

Glen Innes area

This area in the New England district is probably the most productive of sapphires and associated gemstones in New South Wales. The maps of the well-known Frasers Creek, Swan Brook, Horse Gully, Wellingrove Creek, Reddestone Creek and Kings Plains show areas familiar to sapphire-knowledgeable fossickers. The land bounded by imaginary lines drawn between Glen Innes, Armidale and Inverell has been the subject of recent satellite image processing, and pictures show the actual sapphire-bearing rocks to be in the vicinity of the ancient Maybole Volcano south-south-west of Glen Innes. The Department of Minerals and Resources has produced a fine full-colour booklet titled *Sapphires in New*

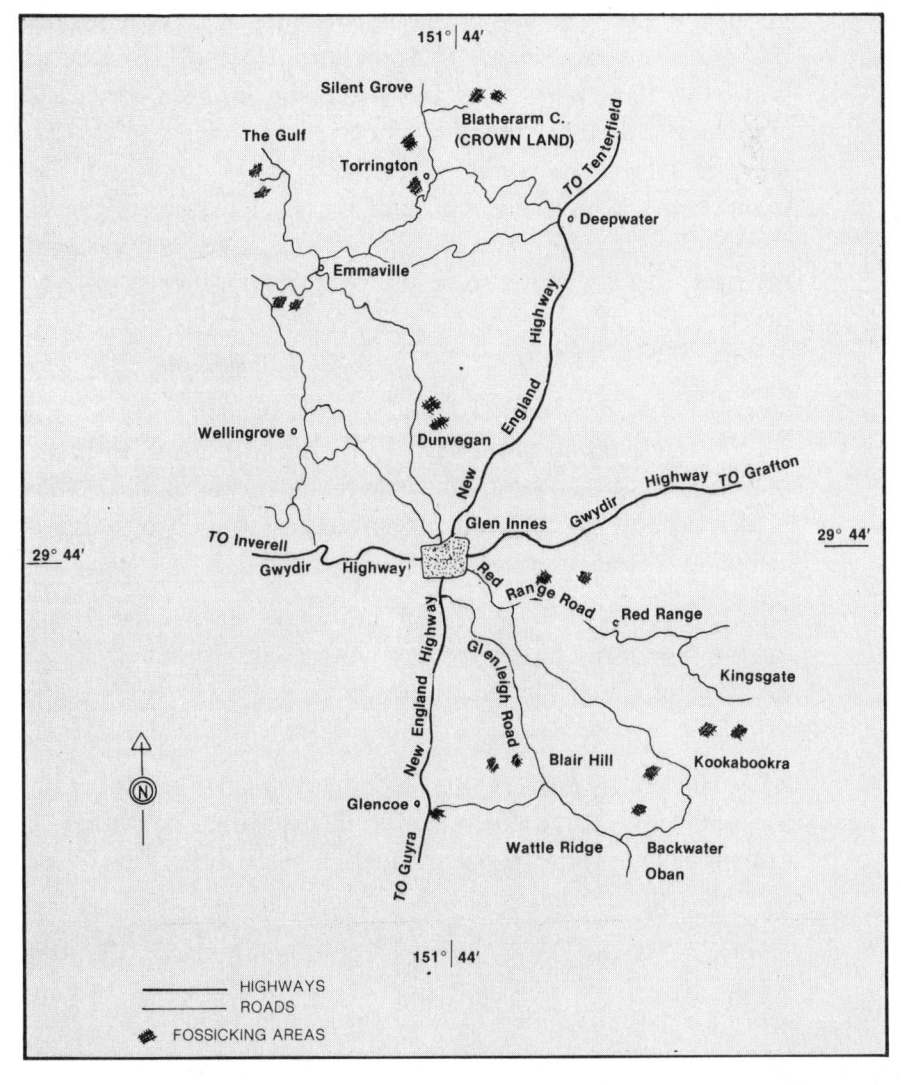

Fossicking areas around Glen Innes, New South Wales. (Not to scale.)

South Wales. This publication is available free of charge and should be in the possession of every would-be fossicker or even the interested amateur sapphire lover. Much information is included, both technical and practical, and the booklet is set out with explanatory maps, diagrams and full-colour satellite photos. We highly recommend that you acquire a copy; it will become part of the fossicker's essential reading. Recently there has been considerable rethinking about the theory of sapphire deposits, and as a consequence the questions of 'where to look and why'—questions in which all fossickers have an everlasting interest—are part explained. All that remains is for the dedicated to do the legwork and locate the final resting places of the sapphires. It has often been suggested, and we are subscribers to the theory, that in spite of all the finds which have been made in the past, there remains to be discovered even more than has been taken so far.

At the time of writing the following fossicking areas are available.

Red Range crown land area, 14 kilometres from Glen Innes on the Red Range Road. On the left side of the road past the bridge sapphires, topaz and zircon may be found. Right bank only.

Glencoe crown land area, 22.5 kilometres from Glen Innes on the New England Highway. On the left side of the road just before the bridge sapphires may be found. Left bank only.

Blair Hill crown land area, 21 kilometres from Glen Innes on the Glen Leigh Road. Turn left beside the 'Super Strawberry'. On the right side of the road just over the bridge look for gems in the right bank.

Kingsgate Mines, 32 kilometres from Glen Innes on the Red Range Road. On the right side of the road quartz crystals may be found. Prior permission is required from the owner. Contact the New South Wales mines department for further information.

Blatherarm Creek crown land area, 20 kilometres past Torrington on the Silent Grove Road. At the 'B' sign turn right and travel for 3 kilometres. Look for aquamarine, topaz and quartz.

Torrington, 65 kilometres from Glen Innes via Deepwater. Old mines abound. It is suggested that intending fossickers check the local hotel for cold beer and local fossicking information. Look for quartz, topaz, beryl and mineral specimens.

Frappell's Topaz Farm, 65 kilometres from Glen Innes via Emmaville, on the Gulf Road. A small fee is charged. Topaz and aquamarine may be found.

Dunvegan Sapphire Company, 13 kilometres from Glen Innes on the Emmaville Road, past the bridge on Reddestone Creek. Fee payable. Sapphires and other gems may be found.

Kookabookra crown land area, 45 kilometres from Glen Innes. Follow the Red Range Road, turn right to Pinkett and continue to the Mitchell River. Past the bridge many of the tracks on the right

A masterpiece in quartz, this large quartz crystal stood for many years outside the courthouse at Glen Innes in northern New South Wales. The crystal weighs in excess of seventy-two kilograms, and is much larger than usually found in this area.

will take fossickers to the creek, where topaz and sapphires may be found.

Wattleridge, 45 kilometres from Glen Innes via Glencoe. Turn left past Mt Mitchell at the sign. Daily fee payable. Sapphires, topaz and quartz crystal may be found here.

Wellingrove, 34 kilometres from Glen Innes. Take the Wellingrove Road, cross Stoney Creek and go through the gate 100 metres on the right. Continue to the creek through the second gate 200 metres on right. Look for sapphires and other gems.

Lightning Ridge

ACCESS

Coach/Rail   A package of seven days' duration is offered by New South Wales Countrylink, (rail/coach tour). It proceeds from Sydney by the XPT train to Dubbo and involves coach from thence to Lightning Ridge. Accommodation and field tours are included in this package. Graham Coaches in Toowoomba, Queensland, also offer three-day tours.
Enquiries on all tours may be made through travel agents or from the Lightning Ridge Tourist Information Centre direct.
Air   There are six flights from Sydney to Lightning Ridge each week made by Hazelton Airlines and Air Link. Single-day return flights direct to 'the Ridge' are available from Challenge Airlines on Queensland's Gold Coast. Enquiries on flights may also be made through travel agents or from the Lightning Ridge Tourist Information Centre.
Road   Lightning Ridge is 764 kilometres from Sydney and 74 kilometres from Walgett. The best route if driving from Sydney is along the Great Western Highway to Bathurst, on to Dubbo via the Mitchell Highway, and then to Gilgandra, Coonamble and Walgett. From Walgett take the Castlereagh Highway, which continues to the Queensland border, to Lightning Ridge. The road from Walgett to Collarenabri and on to Moree is now fully sealed and by the middle of 1997 the road from Melbourne via the Castlereagh Highway right through to Cairns in Far North Queensland will also be in a similar condition.

If driving from the Brisbane area, go to Moree, then to Collarenebri, and then to Lightning Ridge, but only use the road from Collarenebri to Lightning Ridge if the worst of road conditions do not deter you. The alternate route is via Coonabarabran. This covers more kilometres, but the driving conditions are good.

ACCOMMODATION

There are presently four motels in Lightning Ridge: the Black Opal, the Bluey Motel, the Lightning Ridge Motor Village and the Wallangulla

'Painted Lady' opal from Andamooka, South Australia

A 'Yowah Nut' or boulder opal from Yowah, Queensland

Rough opal from the Grawin field near Lightning Ridge, New South Wales

Agate-filled thunder egg from Agate Creek in North Queensland

Half an agate- and crystal-filled thunder egg from Mt. Hay in Queensland

Chrysoprase, a form of chalcedony, coloured with nickel oxide from Queensland

Crocoite crystals from Zeehan in Tasmania. These are quite rare.

Azurite/Malachite from Cobar in New South Wales. This is one of the ores of copper.

This is a golden variety of rutilated quartz found in Tingha in New South Wales.

Two 'fishtail' twin gypsum crystals which are fairly rare

A septarian nodule which is a curiosity of ironstone concretion.

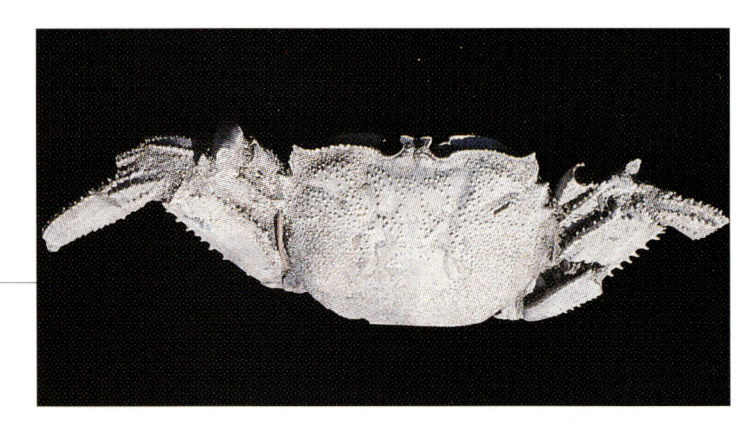

A fossilised crab from Kununurra in the north of Western Australia

An opal 'pineapple' from White Cliffs in New South Wales

Fossilised (petrified) wood complete with ancient worm holes

A very colourful variation of opalised wood from Queensland

Parti-coloured sapphire rough from Inverell in New South Wales

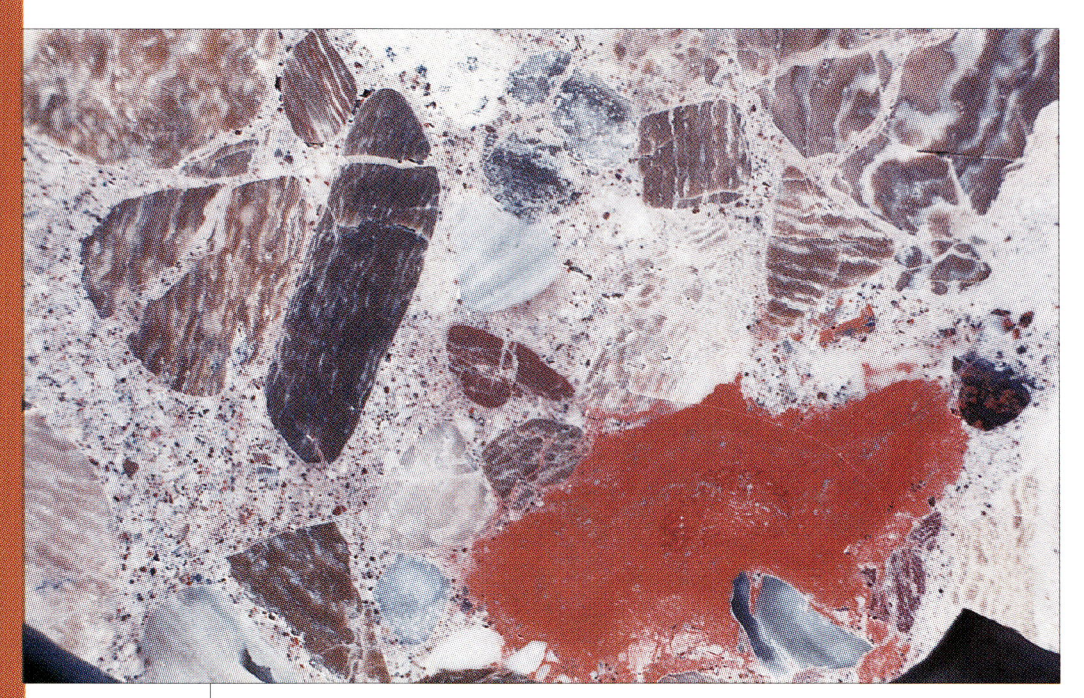

'Fruit salad' agate from Agate Creek in North Queensland

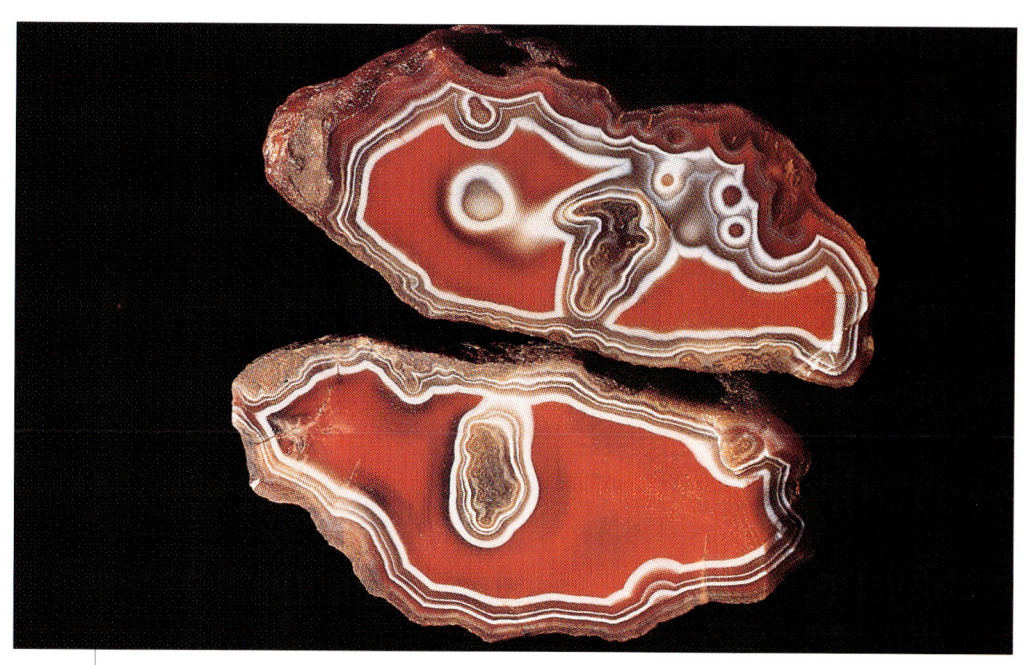

Agate slices from Agate Creek in North Queensland

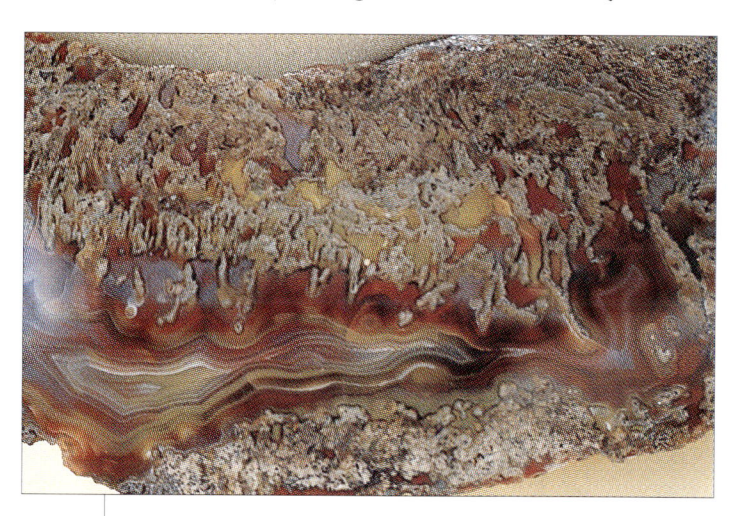

This variety of 'moss agate' contains impurities from when it was formed millions of years ago

A colourful slice of agate from Kangaroo Valley in New South Wales

Zircon crystals with weathered rough from Mud Tank near Harts Range in the Northern Territory

Prehnite from Prospect Quarry in New South Wales

Golden beryl crystals which are very rare

Motel. Three caravan parks also supply camping and cabin accommodation. They are the famous Tram-O-Tel, the Crocodile Caravan Park and the Lightning Ridge Motor Village.

The biggest opal fields in New South Wales are at Lightning Ridge and surrounding districts.

Lightning Ridge is the largest source of quality black opal in the world. Opal is a gem which is always in demand, and fine quality stones command high prices, thousands of dollars per carat. It is no wonder then that the lure of quick riches brings many people to this outback town. But times are changing for Lightning Ridge.

The 'good old days' for fossickers at Lightning Ridge have undergone some modern changes. This used to be a quaint

Opal 'nobbies' from Lightning Ridge, New South Wales. There could be a valuable opal inside one of these.

and appealing western outpost, but with increased population and amenities it is now a town (properly declared so to be) and is experiencing the highest growth rate of any area in its Shire of Walgett.

While the official 1996 population is quoted as approximately 2800 persons, the more correct figure would be somewhere between 7000 and 10 000 if the population in the surrounding opal fields is included. Lightning Ridge's central school has 402 pupils and 34 teachers, many of whom work part-time.

There is every modern convenience to tempt visitors to extend their stay at Lightning Ridge. Motels, hotels, stores, clubs, medical services, galleries, recreational sports

An idle windlass and barricaded shaft entrance at Lightning Ridge, New South Wales. Such abandoned shafts are required to be made safe against accidental entrance. The wire shown here is just one way in which this is achieved.

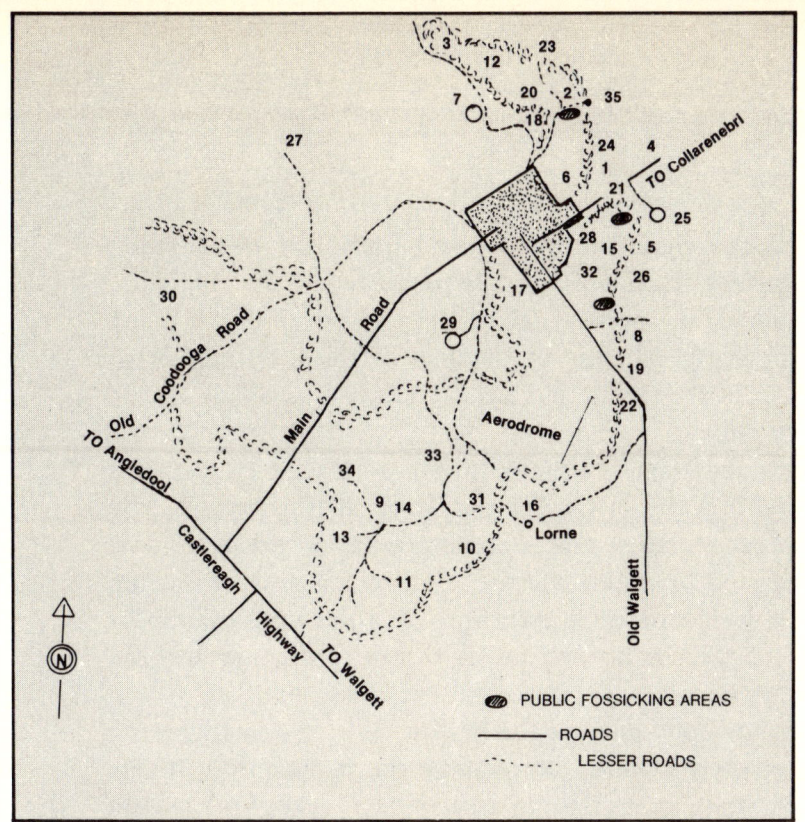

1	Angledool
2	Bald Hill
3	Bill the Boer
4	Bore Baths
5	Butterfly
6	Canfells
7	Cemetery
8	Dry Rush
9	Four Mile
10	Frog Hollow
11	Four Mile Flat
12	Hornet's Rush
13	Hawks Nest
14	Hearts & Spices
15	Kingfisher
16	Lorne Homestead
17	New Town
18	New Chum
19	New Nobby
20	Old Chum
21	Old Town
22	Old Nobby
23	Potch Point
24	Pumpkin Flat
25	Pistol Club
26	Pony Fence
27	Puddling Tanks
28	Sims Hill
29	Sporting Car Club
30	Thorley's Six Mile
31	Three Mile
32	Telephone Line
33	Vertical Bills
34	Deep Belars
35	Walk in Mine

Lightning Ridge opal fields, New South Wales. (Not to scale.)

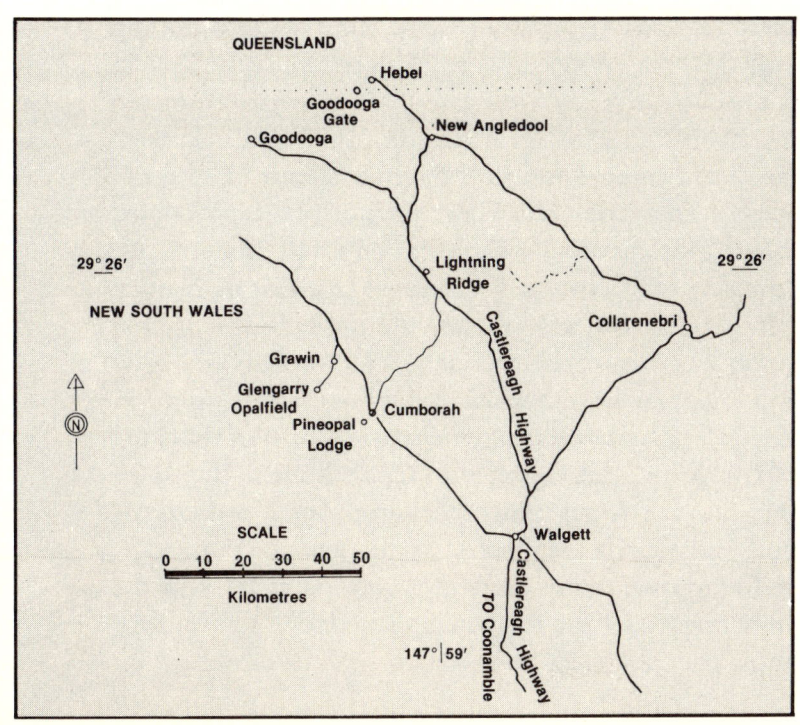

Opal fields centred on Lightning Ridge, New South Wales, with connecting roads shown.

facilities, churches, souvenir shops, laundromat, liquor store, and jewellery stores are all to be found here.

In spite of the presence of modern jewellery shops, many an opal changes hands in the bar of the pub, shown to intending buyers while it nestles on cotton wool in an old cigarette tin.

Lightning Ridge is situated in the northwestern plains of New South Wales in the Walgett Shire. It is surrounded by rich wheat-, cattle- and oilseed-producing areas, and is characterised by its kilometres of glistening pink and white mullock heaps of opal dirt, thrown up by thousands of miners in nearly a century of frantic searching for the precious gem.

April to September are the best months to visit Lightning Ridge, when the winter days will be warm, even if the nights are cold. Take light clothing as well as a jacket for the cool nights. Sturdy footwear and a wide-brimmed hat are also recommended.

Officers of the Lightning Ridge Tourist Information Centre in Morilla Street or the Walgett Shire offices in Walgett are always ready with essential information for the fossicker. Every aspect of your stay in 'the Town' has been anticipated and catered for with many brochures and maps of the whole area available. Phone enquiries are welcomed on (068) 29 1466, as are fax enquiries on (068) 29 0565. We would personally recommend that intending visitors contact the most friendly officers of the Lightning Ridge Tourist Association well before their actual visit comes about.

Fossicking areas have been set aside for visitors, in many cases by operating commercial enterprises. Some of these are the Walk-in-Mine, the Spectrum Opal Mines, the Bush Museum, the Big Opal near the aerodrome, and at the Kangaroo Hill tourist complex. The residents of the Ridge, we have found over many years, are ever ready to assist the 'new chum' visitor or the revisiting 'old hand' and are some of the most friendly people one would wish to meet.

It should be mentioned here for the safety of visitors that they must exercise care when walking or driving about the opal fields, as subsidences may occur at any time where old shafts have collapsed. Especially, keep children under strict supervision, and, of course, do not throw stones down shafts.

White Cliffs

ACCESS
Air Hazelton and Kendall Airlines both service the area from Sydney and Adelaide to Broken Hill and thence onward by coach.
Road Road access is via the Barrier Highway to Wilcannia, 193 kilometres east of Broken Hill.
Coach Silver City Tours conduct excursions of the field.

ACCOMMODATION
White Cliffs White Cliffs Hotel, PJ's Underground Bed and Breakfast, Underground Dugout Motel, Polpah Station Farmstay, Opal Pioneer Reserve (camping with facilities).
Wilcannia Victory Park Caravan Park, Court House Hotel, Wilcannia Motel.

An aerial view showing the mullock heaps of White Cliffs opal field in New South Wales.

White Cliffs is an area in which the fossicker can find precious opal. It is 128 kilometres northwest of Wilcannia. The best time to visit White Cliffs is between April and October.

These remote opal fields were Australia's first; they began giving the world opal as long ago as 1889. White Cliffs is still visited each year by as many people as a decade ago, all with the same keen fossicking instinct.

Regular coach services from Broken Hill to the town have improved markedly over the last few years.

The people of the very durable White Cliffs community will make any visitor very welcome. Fame came to the field a few years ago with the rare find of a large opalised plesiosaur, a member of the ancient fish-like family of animals, believed to exist 120 million years ago. The discovery was made by Ken Howarth and was put on display in White Cliffs for tourists to see. It is still available for viewing from time to time.

Gemstone occurrences

Abercrombie River
  garnet, rock crystal, sapphire
Adjinbilly
  garnet
Airly Mountain
  diamond
Albury
  garnet
Angledool
  precious opal
Ann River
  zircon
Apple Tree Gully
  zircon
Armidale
  topaz, zircon
Ashford
  diamond
Attunga
  garnet

Baan Baa
  agate, jasper, petrified wood, sapphire, thunder eggs
Backwater
  topaz
Bald Nob
  agate, sapphire, topaz, zircon
Ballina
  diamond
Baradine
  agate, jasper, petrified wood
Barraba
  agate, common opal, diamond, garnet, jasper, opalised wood, petrified wood, sapphire, topaz, tourmaline, zircon
Bateman's Bay (south)
  turquoise
Bathurst district
  diamond, garnet, spinel, topaz

Beardy Waters
  garnet, sapphire, topaz
Bellata
  agate, carnelian, jasper, petrified wood
Bendemeer
  chalcedony, diamond, rhodonite, rock crystal, topaz, tourmaline
Ben Lomond
  sapphire, zircon
Berigal Creek
  opalised wood, petrified wood
Berridale
  agate
Berrima
  sapphire, zircon
Bingara
  chalcedony, diamond, garnet, jasper, petrified wood, rock crystal, sapphire, spinel, topaz, tourmaline, zircon
Black Range
  beryl
Blatherarm Creek
  beryl
Blayney
  malachite
Bodalla
  turquoise
Boggabri
  agate, amethyst, chalcedony, common opal, diamond, garnet, jasper, obsidian, petrified wood, thunder eggs, topaz, tourmaline, zircon
Bolivia
  moonstone
Boonoo Boonoo
  topaz
Botolbar
  steatite
Bowling Alley Creek
  fluorite, garnet
Bowning
  agate
Broken Hill
  amethyst, azurite, beryl, fluorite, pyrite, rhodochrosite, rhodonite, sphene, zircon
Bucharoo
  steatite
Bugaldie
  agate, aragonite, jasper, opalised wood, petrified wood
Bulahdelah
  alunite
Bullock Mountain
  garnet, sapphire, topaz
Bundarra
  tourmaline
Bungonia
  beryl
Calula Creek
  diamond
Carcoar
  chalcedony, common opal, garnet, hematite
Coalcliff Beach
  chalcedony, chrysoprase
Cobar
  azurite, chrysocolla, malachite
Cobham
  agate
Collarenebri
  agate, jasper, petrified wood, topaz
Copes Creek
  diamond, sapphire, zircon
Copeton
  diamond, garnet, petrified wood, rhodonite, sapphire, spinel, topaz, zircon
Conona
  garnet
Coonabarabran
  opalised wood
Cow Flat
  beryl
Cowriga
  agate
Crookwell
  beryl, diamond, garnet, ruby, sapphire, topaz, zircon
Cudgegong River
  diamond, sapphire, topaz, zircon
Cumborah
  agate, jasper, petrified wood, topaz

Danglemah
  rhodonite
Delungra
  common opal, diamond
Doon Doon
  agate, chalcedony, thunder eggs
Drake
  agate, azurite, rock crystal
Dubbo
  agate, diamond, chalcedony,
  petrified wood
Duckmaloi River
  sapphire, zircon
Duncan's Creek (Niangla)
  rhodonite
Dundee
  sapphire, topaz
Dungowan
  nephrite
Ebor
  sapphire
Edgeroi
  opalised wood, petrified wood
Elsmore
  beryl, sapphire, smoky quartz,
  zircon
Emmaville
  aquamarine, azurite, beryl,
  citrine, diamond, emerald, rock
  crystal, topaz, tourmaline in
  quartz
Euriowie
  diamond
Fish River
  garnet
Folleys Creek
  common opal
Frasers Creek
  beryl, emerald, sapphire, topaz,
  zircon
Garie Beach
  chalcedony
Giant's Den Knob (near
  Bendemeer)
  actinolite
Gilgai
  cairngorm, citrine, smoky
  quartz, tourmaline in quartz
Girilambone
  malachite

Glencoe
  sapphire
Glen Creek
  beryl, topaz
Glen Eden
  beryl
Glen Elgin
  amethyst, sapphire
Glen Innes
  beryl, cairngorm, citrine,
  fluorite, garnet, rock crystal,
  rutilated quartz, sapphire,
  spinel, topaz, tourmaline
Goonoo Goonoo
  petrified wood
Gorrianawa
  agate, aragonite, opalised
  wood, petrified wood, red
  jasper
Grabben Gullen
  diamond, sapphire
Grafton
  garnet
Grants Lookout
  turquoise
Grawin
  precious opal
Green Hills
  sapphire
Grove Creek
  agate
Gulgong
  chalcedony, garnet, ruby, topaz
Gum Flat
  peridot
Gundagai
  garnet, topaz
Gunnedah
  chalcedony, petrified wood,
  prehnite
Gunningbland
  agate
Guy Fawkes
  jet
Guyra
  emerald, garnet, rock crystal,
  rutilated quartz
Gwydir River
  agate, sapphire

Halls Creek (Moonbi Range)
    rhodonite, rose quartz,
    tourmaline
Hanging Rock
    agate, common opal, garnet,
    jasper, prase, rock crystal,
    sapphire, serpentine, zircon
Hardwicke
    garnet
Harilah
    steatite
Hartley
    garnet
Hill End
    diamond, ruby, sapphire
Hogarth Range
    labradorite
Horse Arm Creek (near Attunga)
    fluorite
Horse Gully
    ruby, sapphire
Howell
    diamond
Hunter River
    chalcedony
Hutchison's Lode
    moonstone
Inverell
    aquamarine, beryl, diamond,
    garnet, jasper, peridot, rock
    crystal, ruby, rutilated quartz,
    sapphire, spinel, topaz, zircon
Kangaroo Flat
    beryl
Kangaroo Valley
    agate
Kempsey
    andalusite, rhodonite
Kiama
    agate, chrysoprase
Kiandra
    beryl, emerald
Kookabookra
    topaz
Kulnura
    aragonite
Lake Macquarie
    petrified wood

Lake Mummuga
    turquoise
Leadville
    petrified wood
Lightning Ridge
    petrified wood, precious opal,
    topaz
Lismore
    agate
Lucknow
    nephrite
Lue
    chalcedony
Macintyre River
    agate
Macquarie River
    diamond, garnet, ruby
Maitland
    chalcedony
Manilla
    aragonite, opalite
Mann River
    sapphire, zircon
Mary-Anne Creek
    sapphire
McGlanglin Creek
    ruby
Milparinka
    petrified wood
Mittagong
    amethyst, diamond, garnet,
    sapphire, topaz
Moama
    garnet
Mole Tableland
    sapphire
Monaro
    agate
Monros Creek
    common opal
Moonbi Range
    rhodonite, rose quartz,
    tourmaline
Moree
    agate
Mount Eltie
    fluorite
Mount Hope
    malachite

Mount Robe
  fluorite
Mount Tennyson
  garnet
Mount Werong
  sapphire
Mount Wingen
  agate
Mudgee
  diamond, garnet, ruby,
  sapphire, spinel, steatite, topaz,
  zircon
Mullumbimby
  hyalite, labradorite
Murrumbidgee River
  citrine
Murrumburrah
  agate, garnet
Murwillumbah
  agate, thunder eggs, turquoise
Namoi River
  sapphire
Nandewar Range
  diamond, petrified wood
Narrabri
  agate, diamond, opalised wood,
  petrified wood
Native Dog Creek
  amethyst, diamond, garnet,
  ruby, sapphire, spinel,
  tourmaline, zircon
Newstead
  amethyst, chalcedony, rock
  crystal, sapphire
Niangla
  rhodonite, sapphire, zircon
Nimmitabel
  ruby
Nora Head
  chalcedony
Nullamanna
  sapphire
Nundle
  agate, chalcedony, garnet,
  jasper, prase, rhodonite, rock
  crystal, sapphire, zircon
Oban
  agate, amethyst, chalcedony,
  citrine, emerald, garnet, rock
  crystal, sapphire, smoky quartz,
  topaz, tourmaline in quartz,
  zircon
Oberon
  amethyst, diamond, garnet,
  ruby, sapphire, spinel,
  tourmaline, zircon
Ophir
  beryl
Pambula
  garnet
Paradise Creek
  beryl, emerald, zircon
Peel River
  rock crystal
Perch Creek
  thunder eggs
Pine Ridge
  diamond
Pond's Creek
  garnet
Poolamacca
  garnet
Port Macquarie
  opalite, rhodonite
Porters Retreat
  sapphire
Prospect
  prehnite
Pyramul Creek
  diamond
Quirindi
  agate, chalcedony, jasper
Reddestone Creek
  garnet, sapphire, topaz
Red Range
  citrine, garnet, tourmaline in
  quartz
Reedy Creek (near Bathurst)
  diamond
Richmond River
  chalcedony
Rockley
  citrine, tourmaline in quartz
Ruby Hill
  garnet
Sapphire
  sapphire, zircon

Scrubby Gully
  beryl
Sheep Station Creek
  fluorite
Shoalhaven River
  diamond, jasper, topaz
Sidmouth
  garnet, tourmaline
Silverton
  garnet, tourmaline
Singleton
  agate
Somerton
  chalcedony
Spring Creek
  common opal
Stanborough
  beryl
Stannum
  beryl, topaz
Stoney Creek
  zircon
Surface Hill
  beryl
Swanville
  chalcedony
Tallong
  garnet
Tambar Springs
  prehnite
Tamworth
  garnet, jasper, rhodonite
Taree
  jasper
Tent Hill
  topaz
Thackeringa
  garnet, kyanite
Therabri
  agate
Tilbuster
  petrified wood, zircon
Tingha
  agate, amethyst, beryl,
  cassiterite, citrine, emerald,
  morion, rhodonite, rutilated
  quartz, sapphire, smoky quartz,
  spinel, topaz, tourmaline in
  quartz

Tintenbar
  common opal, precious opal
Tooraweena
  chalcedony, common opal
Torrington
  amethyst, aquamarine, beryl,
  cairngorm, citrine, emerald,
  morion, rock crystal, smoky
  quartz, topaz, tourmaline,
  tourmaline in quartz
Trunkey Creek
  agate, garnet
Tuena
  rock crystal
Tumbarumba
  beryl, emerald, garnet, sapphire,
  spinel
Tungsten
  topaz
Turon River
  diamond
Tweed River
  agate
Tyalgum
  chalcedony
Unanderra
  aragonite
Upper Tarlo
  diamond
Uralla
  citrine, diamond, garnet, spinel,
  topaz, zircon
Vegetable Creek
  beryl
Walcha
  petrified wood, rhodonite
Wallerawang
  garnet
Wandsworth
  ruby
Warialda
  common opal, garnet, jasper,
  opalised wood, petrified wood,
  sapphire
Warrumbungle Ranges
  prehnite
Washpool Creek
  garnet

Wee Jasper
  garnet, sapphire, zircon
Wee Waa
  agate, carnelian, chalcedony
Wellington
  agate, chalcedony, diamond, ruby
Werris Creek
  carnelian
Wheeo
  diamond, zircon
Whipstick
  garnet
White Cliffs
  precious opal

White Rock (near Drake)
  agate
Wingecarribee River
  diamond, garnet, zircon
Wollongong
  agate, chalcedony, petrified wood
Wunglebung
  beryl
Yass
  garnet
Yetholme
  garnet

Queensland

QUEENSLAND is a very widespread part of the continent with a total area of 1 727 200 square kilometres. The climate varies between tropical and cool temperate, the northern zones being very different from the western and southwestern areas in that they are subject to cyclonic weather in the summer season. The southern border portion of the state near the coast is a place of great variation in temperatures due to elevation above sea level. The northern tableland in the tropical zone also has its spots of temperature extremes again due to elevation, but these are exceptions to the general weather pattern in the northern tableland region.

The coastal strip of the state is primarily volcanic in origin while the majority of the inland areas are located on the Great Artesian Basin and have geological features that have been greatly changed by time. For further detailed information on the geology of the state see *Map 2, Queensland Geology*, obtainable from the Queensland Minerals and Energy Centre for a few dollars.

Thick tropical rainforests cover much of the far northern parts of the state and some of the central and southern mountain areas. The further west one travels, however, vast tracts of dry, inhospitable, desert-like terrain become evident. The inland sea, which covered much of the interior of Queensland millions of years ago and is now completely dried up, has left many areas which are peculiarly Australian in gemstone-producing capability inasmuch as they contain opal. These occurrences are both good grade and commercial. Queensland as a whole is rich in resources not the least of which is minerals, and the gemstone portion of this is both generous and widespread. The most important gemstones in the state are opal and sapphire. There are deposits of sufficient quantity and quality for commercial mining to be economically viable. In fact, it has been said that Australia has two of the most important sapphire deposits in the world in central Queensland at Anakie and northern New South Wales at Inverell. Sufficient be it to say that the world's buyers of sapphire reside more or less permanently in both centres.

The fossicker is looked after to a degree by recent government amendments to the mining act. These provisions are set out in publications which are available from the Department of Minerals and Energy entitled *Guide to*

Treating promising excavated wash on the Anakie gem fields in Queensland. Note the large wooden sieve.

Queensland Fossicking Legislation June 1995 and *Fossicking in Queensland 1996*. These should be studied and understood before planning a fossicking trip to the area. The facts set out in these publications apply to all mining in the state, and if anything is not understood contact should be made with the Department or one of its district offices.

The Department's two brochures mentioned above will be found to be most comprehensive for the fossicker. The *Guide*, dated March 1996, sets out what the law defines as fossicking with all the attending rules, regulations and limitations which are placed upon the fossicker and/or camper. The second leaflet spells out the location of the Designated Fossicking Lands, Fossicking Areas and General

Excavated material from the Sapphire field in Queensland is screened through a coarse sieve to remove all unwanted rocks before treating the finer wash.

The finer dug wash is placed into the willoughbee which is comprised of a sieve over the water tank. The handle, attached to the sieve cradle, is worked up and down in the tank to remove the clay and fine material and to concentrate any sapphire in the wash.

Queensland, showing principal river systems and potential opal-bearing areas.

Permission Areas as set out in the legislation. The gemstones which may be expected in the various areas are clearly listed and a coloured map shows the location of all the Department's Fossicking Areas. Application forms are included showing the scale of fees payable for fossicker's licences and camping permits.

The designated Fossicking Areas as at March 1996 are described further on in this chapter, as is Chinchilla, a designated General Permission Area.

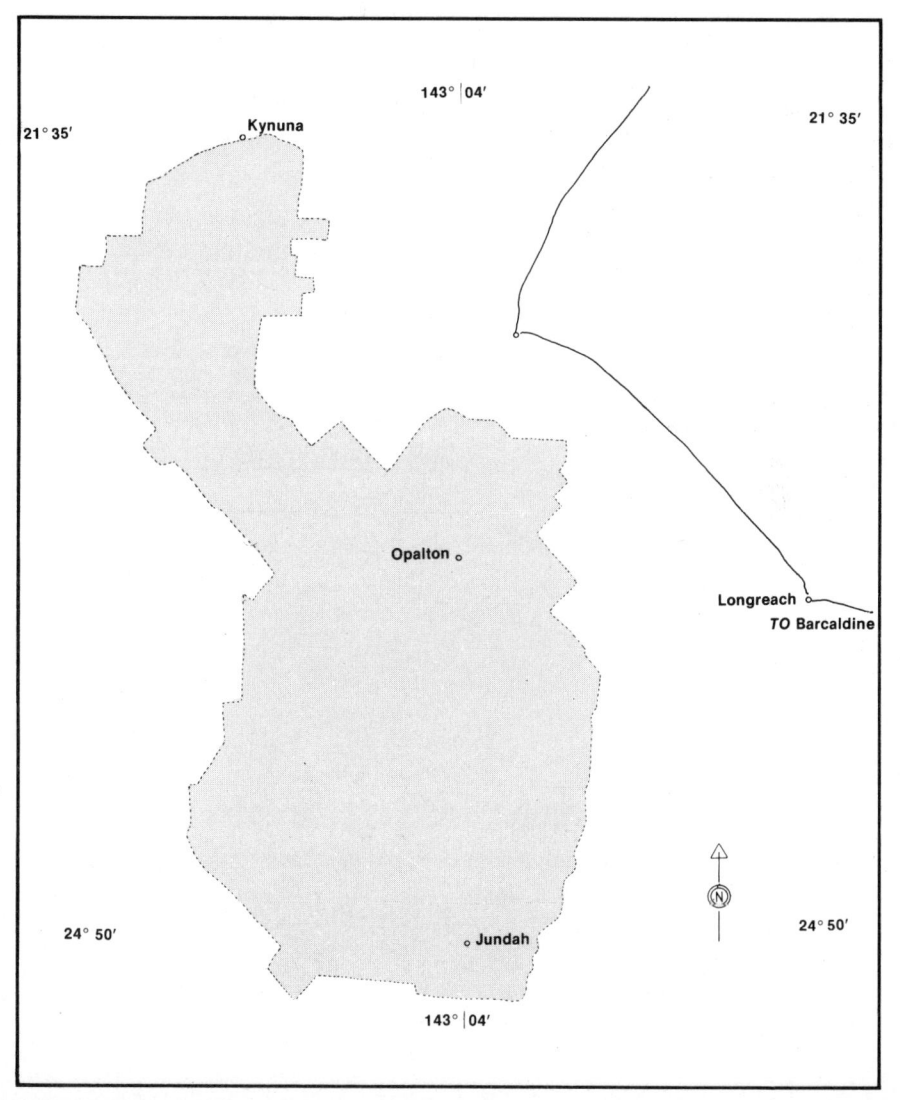

Opal-bearing areas in western Queensland.

An alternative, manual method of treating sapphire concentrate. The larger sieve makes a very strong back essential. (Photo by courtesy of New South Wales Department of Mines and Energy.)

The sieve is upended onto a bag where the sapphires and other gems (if present) are seen and picked out. (Photo by courtesy New South Wales Department of Mines and Energy.)

Roads and safety

Surface roads called 'beef roads' crisscross Queensland in some areas and a great boon they are to the fossicker travelling from point to point over long distances. Unfortunately, many gemstone areas are at the end of unsurfaced or unformed roads and tracks leading off the very convenient access roads. These more minor roads can deteriorate rapidly with the advent of a few millimetres of rain, and unless this is anticipated much anguish can result from the delay or breakdown of vehicle that may result. It's imperative to heed the safety rule of notifying someone before leaving for your destination and estimated time of return.

Queensland is a very large state and the distances which must be traversed to reach a particular gem field make great demands on driver and vehicle. Common sense must be used to determine the daily distance travelled, and good vehicle maintenance is also important to avoid fatigue of people and conveyance. For example, a trip from Brisbane to Agate Creek involves a total of 3600 kilometres at least, there and back, and quite a number of the tracks inland from Forsayth are anything but definite. Hazards in the shape of difficult creek and river crossings in excess of forty in number must be overcome. The vehicle and travellers must be both prepared and able to survive the ordeal in order to fossick for the agate which abounds at the end of the trail. As has been said elsewhere, 'forewarned is forearmed'.

A sapphire crystal section and a coin to show relative size. The hexagonal cross section is typical of the ideal sapphire crystal so much sought by fossickers.

Agate Creek Fossicking Area (Forsayth)

ACCESS

Agate Creek is about 340 kilometres west of Townsville in north Queensland. While suitable for conventional vehicles in normal conditions the roads may present difficulty in the wet season. Tourists can travel from Townsville to Georgetown by fully sealed road, via Charters Towers or Innisfail. Enquiries about road conditions should be made before setting out.

ACCOMMODATION

Camping is prohibited in the Fossicking Area but is catered for nearby.

This area is known for agates of superb colour and pattern; chalcedony and thunder eggs may also be found. Agates can be separated from the decomposed lava formations by hand excavation, and from soil and gravels in or near creeks and gullies (usually dry).

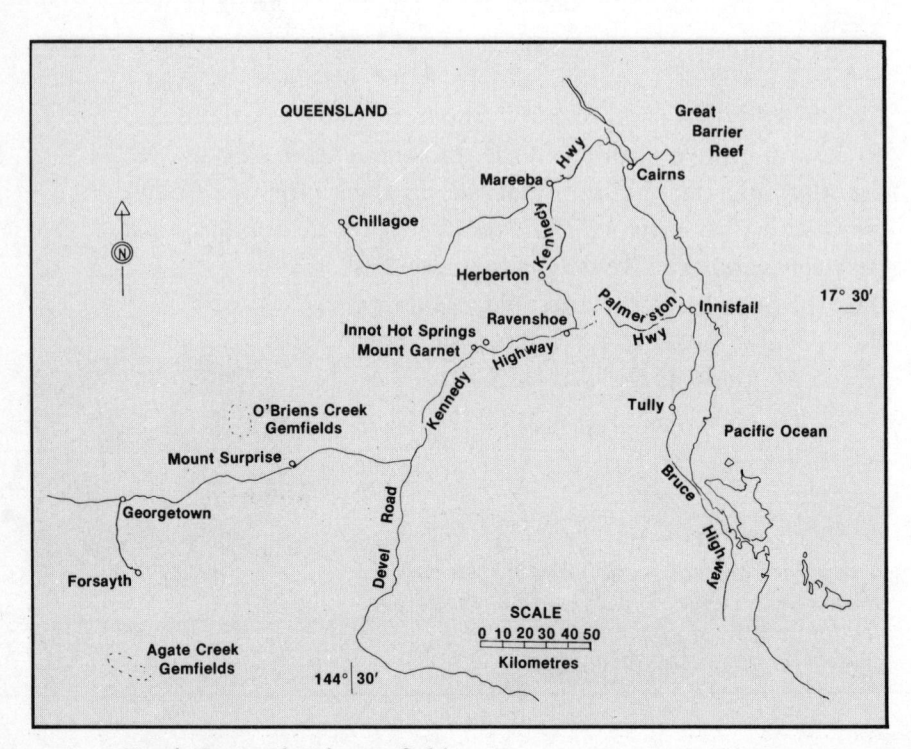

North Queensland gem fields.

QUEENSLAND

Big Bessie Fossicking Area (Sapphire)

ACCESS

Located on the central Queensland gem fields, Big Bessie is a small area which lies within the Sapphire Designated Fossicking Land. Access by rail is via the Rockhampton–Longreach line. Access by road is via the Capricorn Highway.

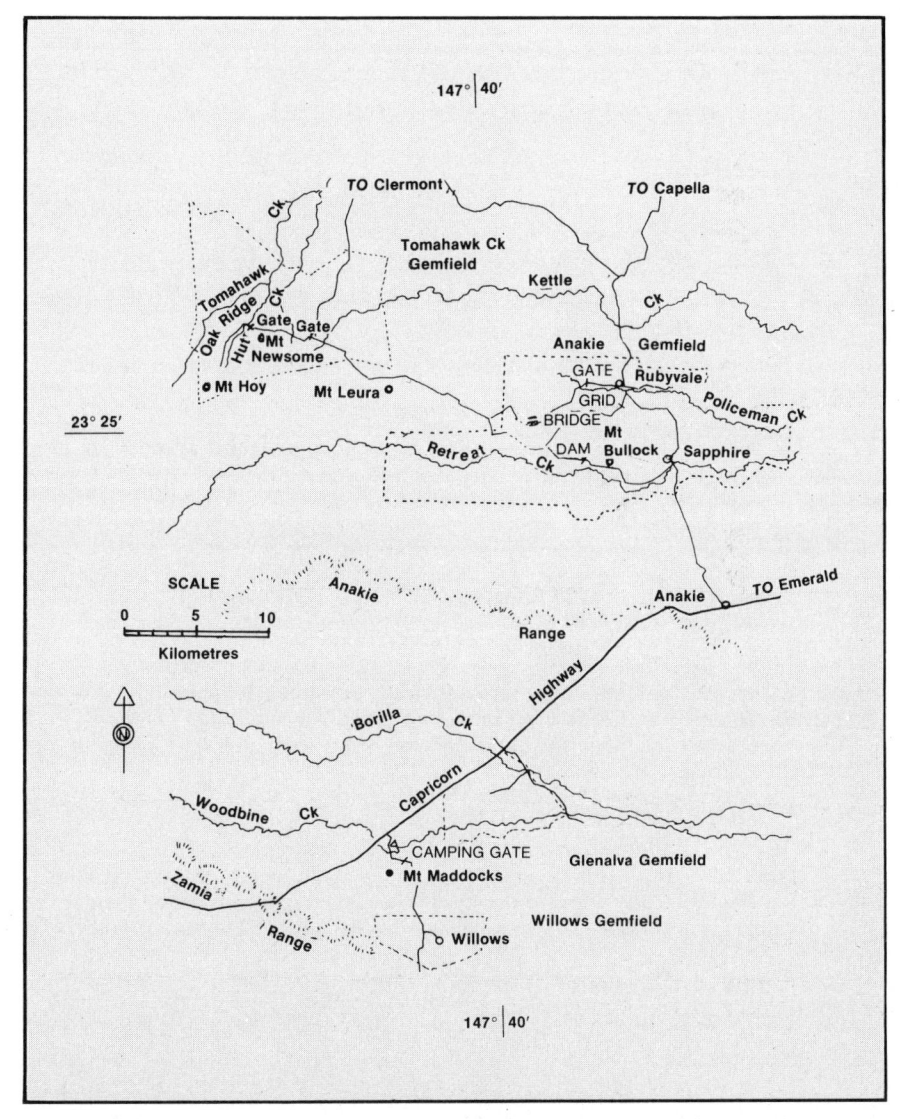

Central Queensland sapphire fields. (Details by courtesy Queensland Government Tourist Bureau.)

ACCOMMODATION

Camping is allowed in the Fossicking Area, for a one-month maximum period. A fossicker's camping permit is required. There is accommodation available at the Anakie Hotel/Motel and at caravan parks at each settlement on the field.

Sapphire-bearing shallow wash covers most of the area and it has been extensively worked. There are still patches of undisturbed ground which could contain sapphires.

Glenalva Fossicking Area (Anakie)

ACCESS

Located on the central Queensland gem fields, Glenalva is about twenty kilometres southwest of Anakie. Access by road is via the Rockhampton–Longreach line. Access by rail is via the Capricorn Highway.

ACCOMMODATION

Camping is allowed for a maximum of one month. A camping permit is required.

A wide variety of coloured sapphires can be found in the shallow wash which covers most of this area.

Graves Hill Fossicking Area (Sapphire)

ACCESS

Graves Hill extends from the southwest corner of the Sapphire Designated Fossicking Land on the central Queensland gem fields. Access by road is via the Rockhampton–Longreach line. Rail access is via the Capricorn Highway.

ACCOMMODATION

Camping is allowed, maximum one month. A camping permit is required. Many people prefer to stay at Emerald where modern facilities are available, and drive to the gem fields as required. There is a wide choice of accommodation available.

Green, yellow and blue sapphires have been found in the shallow wash which covers more than half of this 116-hectare area.

Middle Ridge Fossicking Area (Rubyvale)

ACCESS

Middle Ridge is a small Fossicking Area (thirty-two hectares) located between the Designated Fossicking Lands of Divide and Rubyvale on the central Queensland gem fields. Road access is via the Rockhampton–Longreach line. Rail access is via the Capricorn Highway.

ACCOMMODATION

Camping is allowed for a maximum of one month, a permit being required.

Sapphire and zircon have been found in the mostly shallow wash which covers a good deal of this area.

Tomahawk Creek Fossicking Area (Rubyvale)

ACCESS

Tomahawk Creek Fossicking Area is located approximately twenty kilometres northwest of Rubyvale. Road access is via the Rockhampton–Longreach line. Rail access is via the Capricorn Highway.

ACCOMMODATION

Camping is available for a maximum of one month, a permit being required.

Blue and 'fancy' sapphires may be found in the variable wash in this area.

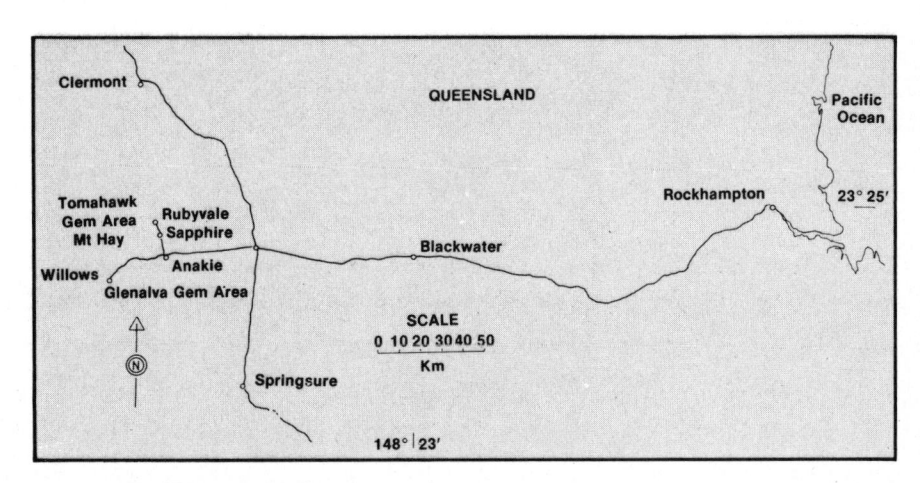

Central Queensland gem fields.

O'Briens Creek Fossicking Area (Mount Surprise)

ACCESS

O'Briens Creek is located northwest of Mount Surprise in north Queensland, about 200 kilometres southwest of Cairns. It may be reached by good sealed road from Cairns but the gem area might need the facility of a four-wheel drive.

ACCOMMODATION

Camping is prohibited in the Fossicking Area. However, camping facilities are available nearby and accommodation is available at the Georgetown Hotel, fifty-nine kilometres east of Mount Surprise.

The main gem variety found here is topaz. Citrine, smoky quartz and aquamarine are also found in the gravels.

Yowah Fossicking Area (Cunnamulla)

ACCESS

Yowah is about eighty-seven kilometres from Eulo which is in turn sixty-five kilometres west of Cunnamulla in the southwest corner of

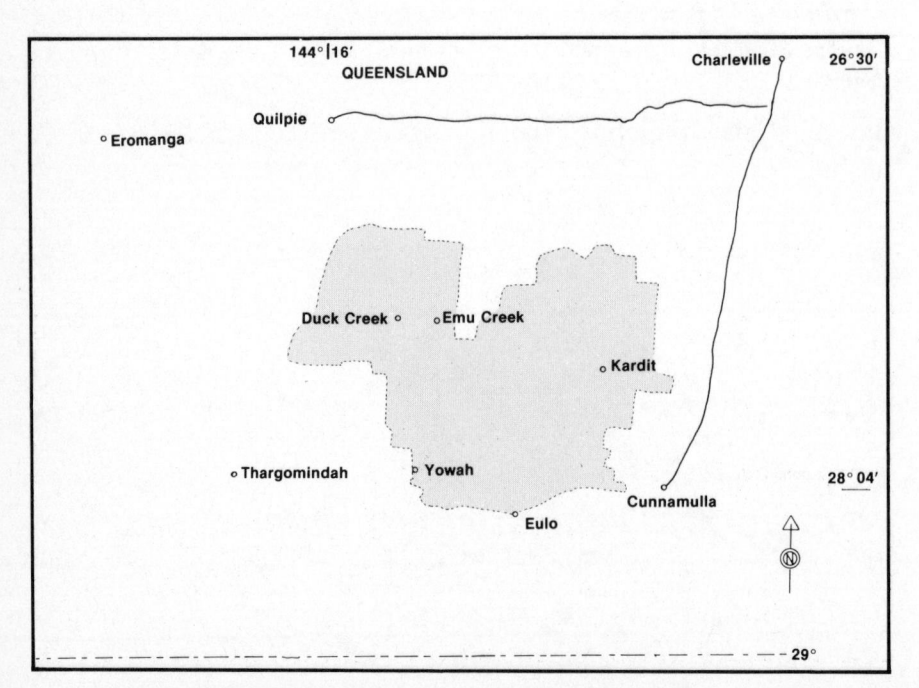

Opal-bearing areas in southwestern Queensland.

Queensland. The most direct road route from southern states is via the trunk highways to Bourke in New South Wales, thence to Cunnamulla and Eulo. Cunnamulla is at the junction of the Balonne and Mitchell Highways. It may also be reached by rail via Charleville from Brisbane.

ACCOMMODATION

Camping is prohibited in the Fossicking Area but camping facilities are available nearby and accommodation is available at the historic Eulo Hotel.

In this region, boulder opal can be found in silicious ironstone nodules, or 'Yowah nuts' as they are known. Chips of opal or ironstone matrix can also be found on the surface or in shallow diggings.

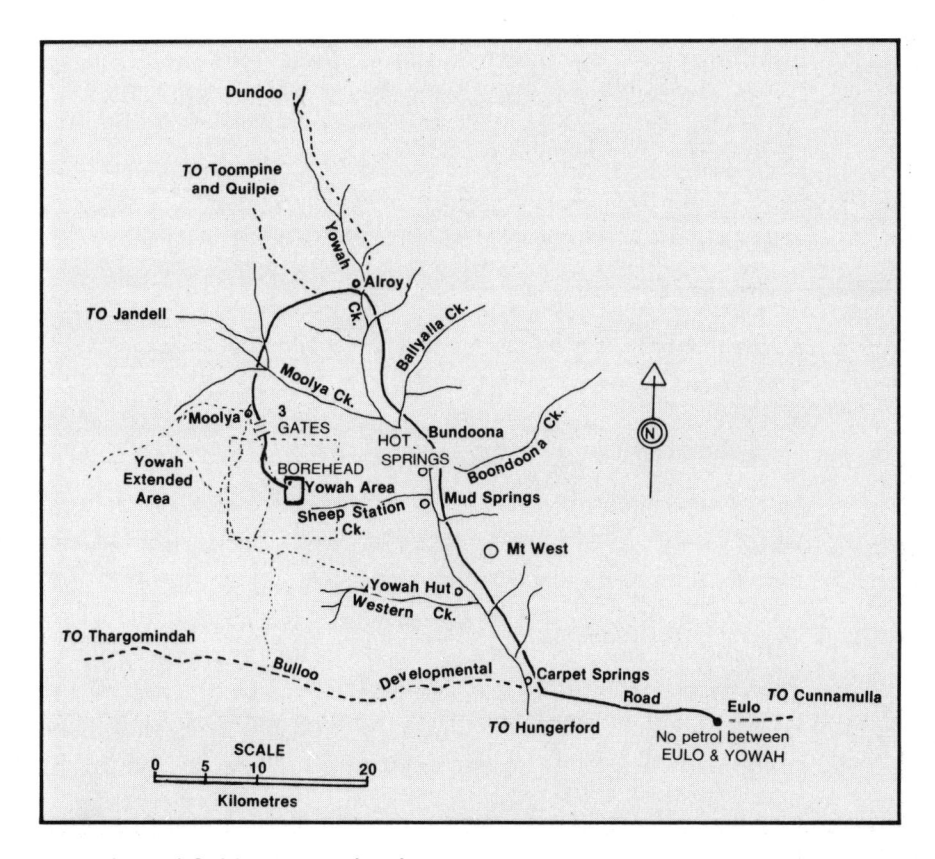

Yowah opal fields, Queensland.

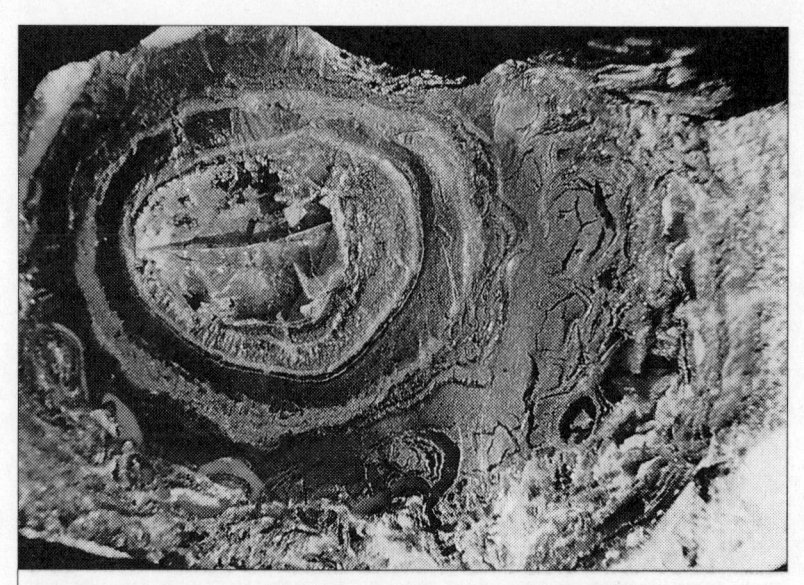

Yowah nut section. Opal forms inside these ironstone concretions and is often of the precious type. A kernel of highly coloured crystal opal is amongst the most beautiful in the world and the most valuable.

Chinchilla

ACCESS

Access to Chinchilla is by coach or car from Brisbane or Toowoomba.

ACCOMMODATION

Accommodation is available at the Chinchilla Hotel/Motel and also at Miles, forty-six kilometres west.

Situated approximately 250 kilometres west-north-west from Brisbane, Chinchilla is in the Brisbane Mining District and is defined as a General Permission Area. Fine petrified wood has been found in the general area over many years and the supply never seems to dwindle. Swipers Gully at Stanthorpe is also in the General Permission Area of the Brisbane Mining District. Phone or fax enquiries should be made from the Department regarding the special conditions which apply to fossicking in Chinchilla and in Swipers Gully.

Gemstone occurrences

Agate Creek
 agate, chalcedony, jasper, thunder eggs, zircon
Almaden
 fluorite, sapphire, topaz
Anakie
 amethyst, chalcedony, diamond, sapphire, topaz, tourmaline, zircon
Atherton Tableland
 peridot
Ballendeen
 topaz
Barkly Tableland
 ribbonstone
Beallah
 zircon
Beatrice River
 sapphire
Beechmont
 amethyst, thunder eggs
Bell
 moonstone
Blackall – Listowel Downs
 opal
Blackbutt
 garnet, sapphire, topaz, zircon
Black Gate
 precious opal
Blackwater
 common opal, petrified wood
Boondoomlia Station (west of Proston)
 tourmaline
Booreeco Creek
 garnet
Brigooda
 garnet
Broadwater Creek
 beryl, topaz
Broken River
 garnet, zircon
Buderim Mountain
 common opal, petrified wood
Bullock Creek
 fluorite
Bulls Creek
 precious opal
Bunerba
 garnet, sapphire, zircon
Calliope
 agate
Camooweal
 ribbonstone
Cattle Creek
 garnet
Cedar Creek
 agate, common opal
Central Creek
 sapphire
Cheviot Hills
 peridot
Chillagoe
 agate, petrified wood, topaz
Chinchilla
 common opal, petrified wood
Chuddleigh Park
 sapphire
Clermont
 azurite
Cloncurry
 azurite, garnet
Cloyna
 agate, amethyst
Coolgarra
 topaz
Coonavilla
 precious opal
Cooper Creek
 common opal
Cooram
 common opal, hyalite
Cordalba
 common opal
Croydon
 agate
Dargalong
 fluorite
Dayboro
 opal
Decloath Creek
 beryl
Diglum Creek
 garnet
Drummond Range
 agate, sapphire

Dry Creek (Newellton)
  topaz
Duck Creek
  precious opal
Elbow
  precious opal
Emu Creek
  precious opal
Emuford
  fluorite, garnet, topaz
Emu Park
  turquoise
Etheridge
  agate, topaz
Eungella
  garnet, zircon
Fiery Cross
  precious opal
Fischerton
  rock crystal, rose quartz, zircon
Galah Creek
  beryl
Georgetown
  zircon
Gilberton
  agate, topaz
Gilbert River
  sapphire
Glassford Creek
  garnet
Glenalva
  diamond, tourmaline
Glenbar
  garnet
Glendarriwell
  sapphire
Goomburra
  sapphire, zircon
Greenvale
  moonstone
Grey Range
  precious opal
Gympie
  rock crystal
Hazeldean
  garnet, zircon
Henrietta Creek
  sapphire, topaz, tourmaline, zircon

Herberton
  agate, diamond, fluorite, sapphire, topaz, zircon
Hunts Creek
  beryl, topaz
Hut Creek
  sapphire
Ingham
  topaz
Innisfail
  garnet
Innot Hot Springs
  topaz
Jordan Creek
  garnet, sapphire, topaz, tourmaline, zircon
Jordan Mining Field
  garnet, topaz
Kettle Creek
  sapphire
Kettle Swamp Creek
  beryl, topaz, zircon
Kilkivan
  chrysoprase
Kingar Creek
  amethyst
Koroit
  precious opal
Kuridala
  amethyst
Kynuna
  precious opal
Lappa
  topaz
Listowel Downs
  opal
Little River
  agate
Lode Creek
  beryl, topaz
Logan River
  amethyst
Longreach
  agate
Lowmead
  amethyst
Lowood
  garnet

QUEENSLAND

Main Range (Toowoomba)
  peridot
Marlborough
  chrysoprase
McCord's (north of Fischerton)
  fluorite
Medway Creek
  sapphire
Mica Creek
  beryl
Minerva Creek
  hyalite, moonstone
Mitchell River
  agate, garnet
Morinish
  garnet
Mount Britten
  common opal
Mount Garnet
  fluorite, garnet, topaz
Mount Grim
  garnet
Mount Hay
  agate, thunder eggs
Mount Hoy
  sapphire
Mount Isa
  andalusite, azurite, chiastolite,
  chrysocolla, citrine, garnet,
  malachite, smoky quartz,
  tourmaline, tourmaline in quartz
Mount Surprise
  garnet, topaz
Mungana
  fluorite
Munna
  garnet
Murgon
  agate, chalcedony
Murong
  chalcedony
Nanango
  agate, sapphire
Nerang River
  agate, chalcedony
Nettle Creek
  topaz
Newcastle Range
  garnet

Newellton
  garnet, topaz
New Rush
  sapphire
O'Briens Creek
  aquamarine, peridot, topaz
Opalton
  precious opal
Opera Creek
  garnet
Peak Downs
  azurite
Pelican Creek
  common opal
Percy River
  garnet
Percyville
  agate, chalcedony
Pikedale
  topaz
Pine Mountain
  common opal
Pinevale
  garnet, zircon
Porters Gap
  moonstone
Proston
  agate, garnet, tourmaline
Quartpot Creek
  aquamarine, beryl, precious
  opal, topaz
Quartz Hill
  beryl, garnet, topaz
Quilpie
  precious opal
Redcap Creek
  garnet
Redcliffe
  agate
Retreat Creek
  sapphire
Rocky Creek
  amethyst, topaz, zircon
Rubyvale
  diamond, sapphire, zircon
Russell Goldfields
  diamond, sapphire, zircon

Sapphire
    amethyst, diamond, sapphire,
    tourmaline
Scotland Hills
    common opal
Serpentine Creek
    sapphire
Seven Mile Diggings (Nanango)
    topaz
Sheep Station Creek
    precious opal
Silver Valley
    garnet, topaz
Spring Bluff
    peridot
Spring Creek
    garnet, sapphire, topaz
Springsure
    labradorite, moonstone, opal
Square Top Mountain
    hyalite
Stanthorpe
    amethyst, cassiterite, diamond,
    rock crystal, sapphire
Stanwell
    common opal
Sugarloaf Creek
    beryl, topaz
Swanfels Valley
    common opal
Swipers Gully
    topaz
Tamborine Mountain
    thunder eggs

Tarampa
    garnet
Thulimbah
    topaz
Tomahawk Creek
    amethyst, diamond, sapphire,
    tourmaline
Toompine
    precious opal
Toowoomba (Main Range)
    peridot
Upper Cattle Creek
    garnet
Warwick
    rhodonite
Watsonville
    garnet, tourmaline
Williams
    agate, amethyst
Willows
    diamond, sapphire, tourmaline
Windera
    agate, amethyst
Winton
    precious opal
Withersfield
    sapphire
Wyberba
    topaz
Yorkys
    garnet
Yowah
    precious opal

Victoria

ALTHOUGH VICTORIA DOES not possess big gem deposits in commercial quantities, there are plenty of gems to interest the amateur fossicker.

The Victorian Department of Natural Resources and Environment has issued a thirteen-page collection of information sheets concerning fossicking, or more correctly prospecting as the entire publication is directed at the recovery of gold, which of course is the most significantly mined mineral in the state. Each sheet is in the form of a question and answer document which covers the legal requirements of the relevant Act as applies to 'hobbyists, fossickers and gemstone seekers', as the Department defines that very large fraternity.

The regulations require that all who wish to fossick must have a Miner's Right which has a currency of up to two years. It costs eighteen dollars per adult (eighteen and over) for the full two years or any part thereof. There is no concession rate.

There are no nominated fossicking areas designated in Victoria. The regulations as they stand are directed primarily at the gold seeker.

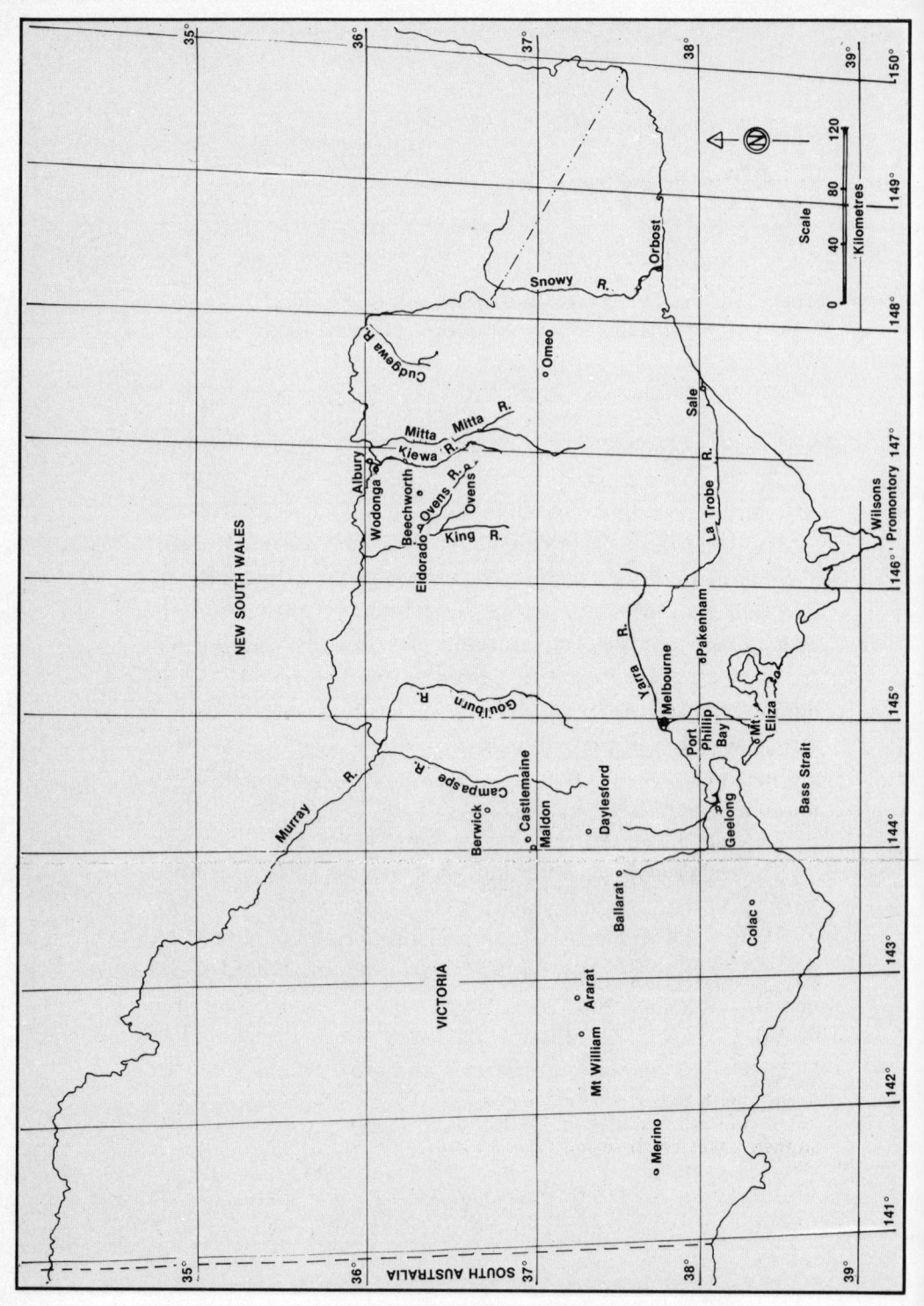

Principal rivers and gemstone-fossicking areas of Victoria.

Beechworth district

ACCESS

Road   From Melbourne, take the Hume Highway to Wangaratta, then the Ovens Highway and the Beechworth Road. From Sydney, take the Hume Highway.
Rail   Take the Melbourne–Albury line to Wangaratta, then bus to Beechworth.

ACCOMMODATION

Facilities for accommodation are very good, with hotels, caravan parks and camping services available. The climate is pleasant in summer, but very cold in winter.

The Beechworth district, 270 kilometres northeast of Melbourne, in particular is a favourite haunt of the gem hunter. Quality stones are not available in abundance, so diligence and patience are necessary.

Beechworth has a colourful history of gold recovery, and in the search for gold other gems have been found, including diamond, citrine, amethyst, agate, morion, jasper, topaz, spinel and zircon. Pyrope garnet occurs at Koetong, seventy kilometres away.

Buchan district

ACCESS

Road   Take the Princes Highway to Bairnsdale, the Omeo Highway to Bruthen, then on to the Buchan Road.
Rail   From Melbourne, take the Orbost line to Bruthen, then road to Buchan.

ACCOMMODATION

Stonehenge Caravan and Camping Park provides comfortable amenities. Accommodation is also available in hotels, motels and camping areas at Bruthen, Lakes Entrance and Nowa Nowa.

The Buchan district, 360 kilometres east-north-east of Melbourne, is known for thunder eggs, jasper, aragonite and calcite.

Heathcote

ACCESS

Road   Access by road is via the Hume Highway to Kilmore, then the Northern Highway to Heathcote.
Coach   Daily coach services are available from Melbourne.

ACCOMMODATION

Motel, hotel/motels, B&Bs, farm cottages and caravan parks are available in and close to town. They vary in quality and price.

Red jasper has been collected at Heathcote, 108 kilometres north-north-west of Melbourne.

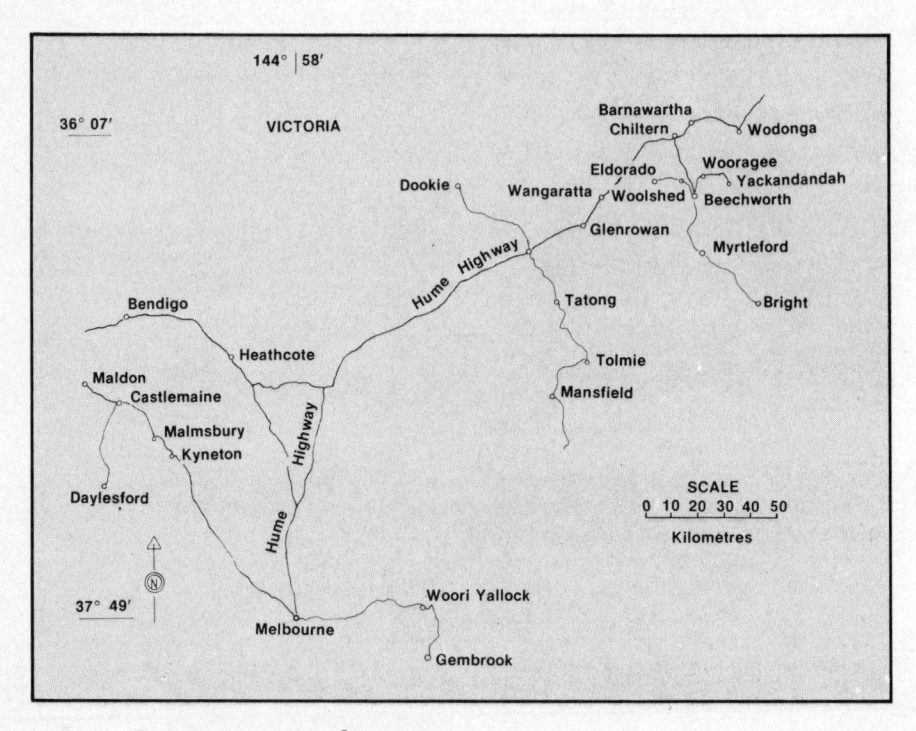

Gemstone areas of Victoria.

Castlemaine district

ACCESS

Road By car take the Midland and Pyrenees Highways.
Rail/Coach Daily rail and coach services available from Melbourne.

ACCOMMODATION

Several quality hotels and accommodation houses as well as two high-grade caravan parks are available in close proximity within the town.

Another old gold area which has a variety of gemstone material to interest the fossicker is the Castlemaine district. The gem areas are within easy distances of Castlemaine itself, and include Fryerstown, Maldon, Harcourt, Heathcote, Trentham, Morrisons, Carisbrook and Mt Alexander.

The following list gives gem localities near Castlemaine and the gems which have been found: Diamond Gully—spinel, quartz, quartz crystals; Fryerstown Road—quartz crystals; Harcourt—cairngorm, citrine, feldspar, topaz, tourmaline, quartz; Jim Crow Creek—cairngorm, citrine, garnet, quartz, sapphire, spinel, zircon; Loddon River—sapphire, spinel, quartz, zircon; Maldon—quartz crystals; Metcalfe—jasper, quartz, sapphire, spinel, topaz, zircon; Middleton's Creek—garnet, sapphire, spinel, zircon; Mt Franklin—feldspar, olivine (peridot or chrysolite), moonstone; Vaughan—diamond, quartz crystals, sapphire, spinel, topaz, zircon; Yandoit—beryl, petrified wood, spinel, topaz, tourmaline, zircon.

Gemstone occurrences

Aberfeldy
  zircon
Ararat
  citrine, garnet, sapphire, topaz
Bacchus Marsh
  common opal, topaz
Ballarat
  garnet, petrified wood, sapphire, zircon
Baringhup
  common opal, hyalite

Barnawartha
  garnet
Bass River
  common opal, petrified wood
Beechworth
  agate, amethyst, beryl,
  chalcedony, chrysoberyl, citrine,
  common opal, diamond, garnet,
  opalite, ruby, sapphire, smoky
  quartz, spinel, topaz,
  tourmaline, zircon
Beenak
  ruby, topaz
Bendigo
  amethyst, peridot, zircon
Berwick
  amethyst, garnet, ruby, sapphire
Blackwood
  garnet, sapphire, zircon
Buchan district
  agate, psilomelane, thunder
  eggs
Bunyip River
  topaz
Camperdown
  peridot
Cape Otway
  chalcedony, jasper
Cardinia Creek
  amethyst, common opal,
  opalite, sapphire
Carisbrook
  topaz
Cassilis
  tourmaline
Casterton
  agate, chalcedony
Castlemaine
  agate, amethyst, diamond,
  feldspar, garnet, jasper,
  moonstone, petrified wood,
  rock crystal, ruby, sapphire,
  spinel, topaz, tourmaline, zircon
Chiltern
  diamond, garnet, sapphire,
  zircon
Colac
  sapphire, zircon

Collingwood
  aragonite
Dandenong
  tourmaline
Daylesford
  amethyst, common opal,
  emerald, peridot, petrified
  wood, ruby, sapphire, zircon
Dereel
  sapphire
Derrinal
  agate
Derrinallum
  agate
Donnellys Creek
  emerald, sapphire, topaz, zircon
Dookie
  chalcedony, jasper
Dundly
  topaz
Dunolly
  topaz
Edi
  turquoise
Eldorado
  agate, amethyst, citrine,
  diamond, garnet, sapphire,
  topaz
Foster
  ruby, sapphire, topaz, zircon
Gelantipy
  common opal
Gellibrand River
  agate, chalcedony
Gembrook
  sapphire, topaz, zircon
Gisborne
  hyalite
Glendinning
  garnet, ruby, sapphire, topaz
Glenrowan
  agate
Goulburn River
  chalcedony, jasper
Grampian Mountains
  common opal
Harrow
  garnet

Heathcote
  chalcedony, jasper
Keilor
  common opal
Kevington
  steatite
Kyneton
  common opal, hyalite
Lake Boga
  smoky quartz
Lal Lal
  topaz, zircon
Leongatha
  sapphire, zircon
Lilydale
  garnet, zircon
Linton
  amethyst
Longford
  garnet
Maldon
  amethyst, beryl, cairngorm,
  citrine, emerald, garnet, peridot,
  rock crystal, topaz, tourmaline
Malmsbury
  common opal, hyalite
Mansfield
  diamond, turquoise
Merino
  topaz, zircon
Mooloort
  agate, chalcedony
Moonlight Head
  agate, chalcedony
Mornington Peninsula
  common opal, ruby, sapphire
Moroka River
  agate, chalcedony
Morwell
  marcasite
Mount Eliza
  ruby, sapphire
Mount Lookout
  peridot, sapphire
Mount Stavely
  common opal
Mount William
  amethyst

Mt Alexander
  tourmaline
Murray River
  agate, chalcedony, jasper
Myrrhee
  turquoise
Northwestern Victoria
  australite (tektite)
Nowa Nowa
  chalcedony, jasper
Omeo
  topaz, zircon
Ovens River
  agate, amethyst, chalcedony,
  garnet
Pakenham
  beryl, ruby, sapphire, topaz
Phillip Island
  agate, chalcedony
Point Leo
  garnet, sapphire, zircon
Reedy Creek
  garnet
Saint Arnaud
  tourmaline
Sassafras Creek
  common opal
Snowy River (lower reaches)
  thunder eggs
Stawell
  garnet, topaz
Strathbogie Ranges
  smoky quartz, tourmaline
Sunbury
  common opal
Talbot
  topaz, zircon
Tallangatta
  garnet
Tanjil
  ruby
Tarrengower
  smoky quartz, topaz
Tatong
  chalcedony, jasper
Tolmie
  sapphire, zircon

Toombullup
  garnet, ruby, sapphire, zircon
Toora
  ruby, sapphire, topaz, zircon
Tubbarubba
  garnet, ruby, sapphire, zircon
Waratah
  topaz, tourmaline
Wellington River
  common opal
Whitfield
  amethyst, turquoise
Wilson's Promontory
  tourmaline
Wodonga
  garnet
Woolshed Creek
  garnet
Wooragee
  chrysoberyl, diamond, zircon
Woori Yallock
  common opal
Yackandandah
  topaz
Yanakie
  garnet
Yandoit
  chalcedony
Yarra River (upper reaches)
  agate, chalcedony, emerald, sapphire, smoky quartz, topaz, tourmaline

South Australia

Because South Australia supplies over ninety-five per cent of the world's commercial opal from its fields at Coober Pedy, Andamooka and Mintabie, the other mineral wealth of this state tends to be overshadowed.

A great variety of gemstone material has been found in South Australia, albeit in small quantities and without commercial possibilities, but nevertheless attractive to the amateur fossicker. Gems found include actinolite, agate, alunite, amazonite, amethyst, andalusite, apatite, aragonite, azurite, beryl, cacoxenite, chrysocolla, chrysolite, chrysoprase, corundum, crystal quartz, diamond, epidote, feldspar, fluorite, garnet, hematite, hyalite, jade, jasper, lazulite, malachite, moss agate, nephrite, olivine, opalite, orthoclase, pseudomalachite, pyrite, smoky quartz, sphene, tourmaline and turquoise.

The Mount Painter–Arkaroola area is one of the most richly endowed gem and mineral regions of South Australia, and is a paradise for fossickers. There is a great diversity of interesting specimens to be located, but fossickers are restricted to taking two kilograms of small specimens and one large specimen. Amethyst, azurite, crystal quartz, feldspar, garnet, jasper, malachite, topaz, tourmaline, turquoise and zircon have been found in this rugged and beautiful locality.

The Yorke Peninsula has yielded interesting mineral specimens and also malachite and azurite. Cowell is now the centre of a commercial venture which is developing quite large deposits of nephrite jade for export. Black material of very high grade has attracted cutters from overseas, particularly the east.

On Kangaroo Island rubellite and indicolite, members of the tourmaline group, have been found.

However, without doubt, opal is the glamour stone of South Australia, and it is to the opal fields that gemstone fossickers sooner or later find their way.

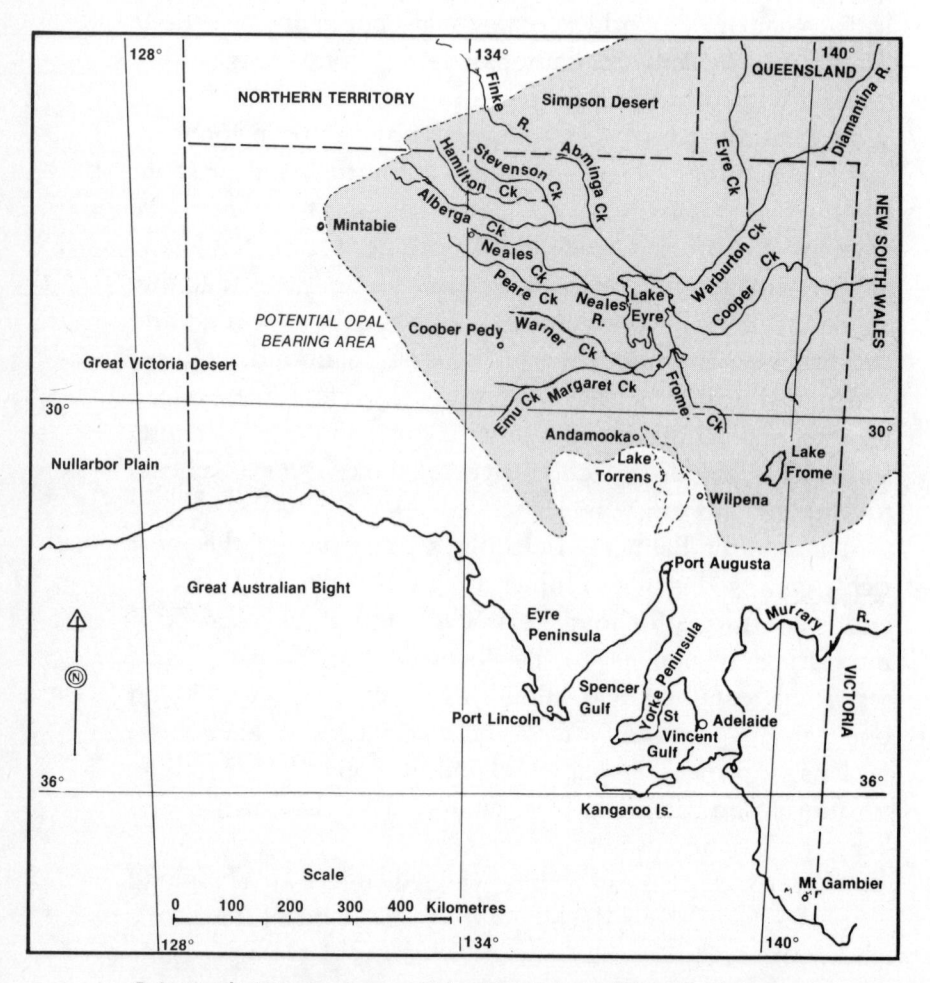

Principal river systems and potential opal-bearing areas of South Australia.

Andamooka opal fields

ACCESS

Road Access to the town is easy along the new fully sealed road from Woomera or from Roxby Downs, thirty-four kilometres away.
Coach Stateliner coaches call on a regular schedule from Adelaide via Woomera and Roxby Downs throughout the week.
Air Air access to Andamooka is by a regular service out of Adelaide with Kendall Airlines.
Tours Kendall Airlines can arrange special tours throughout the area in cooperation with Olympic Dam Tours. A total coverage of the Andamooka country is available with Opal Mine Tours, and special tours of the Woomera Rocket Range, the Andamooka Cattle Station and the Roxby Downs Mine are also available.

ACCOMMODATION

Two caravan parks, plus the Opal Hotel, the Opal Motel and the Duke's Guesthouse accommodate visitors and provide all mod cons.

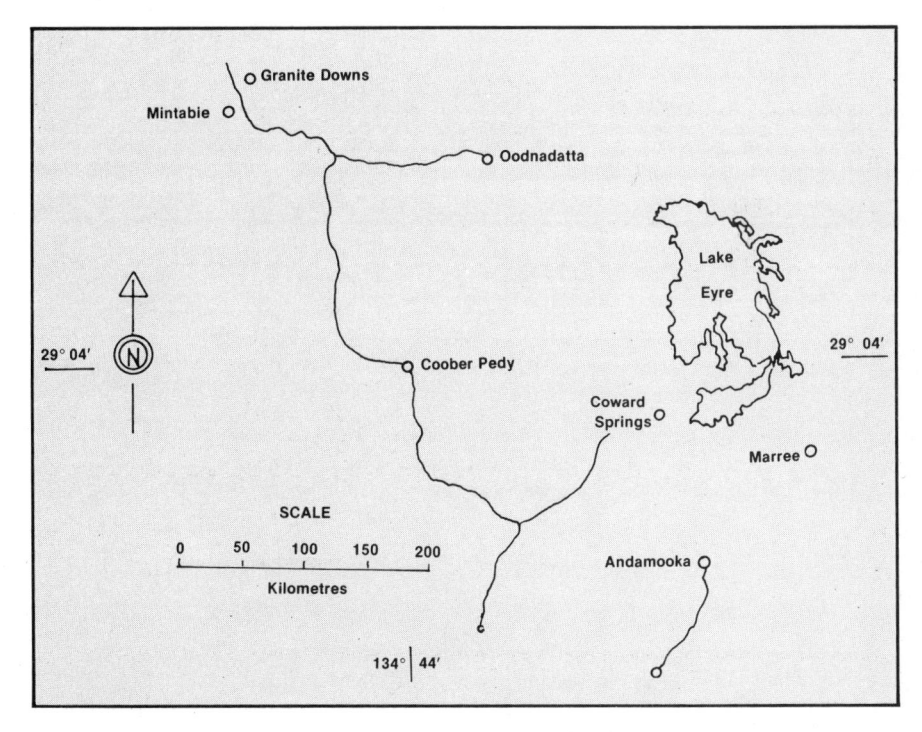

South Australian opal fields, with access roads shown.

Opal was first discovered at Andamooka in 1930, west of Lake Torrens. This desert outpost lies in arid country, on a dry claypan, and has a rainfall of only a few millimetres yearly. The field is 135 kilometres by road north of Pimba and 611 kilometres from Adelaide.

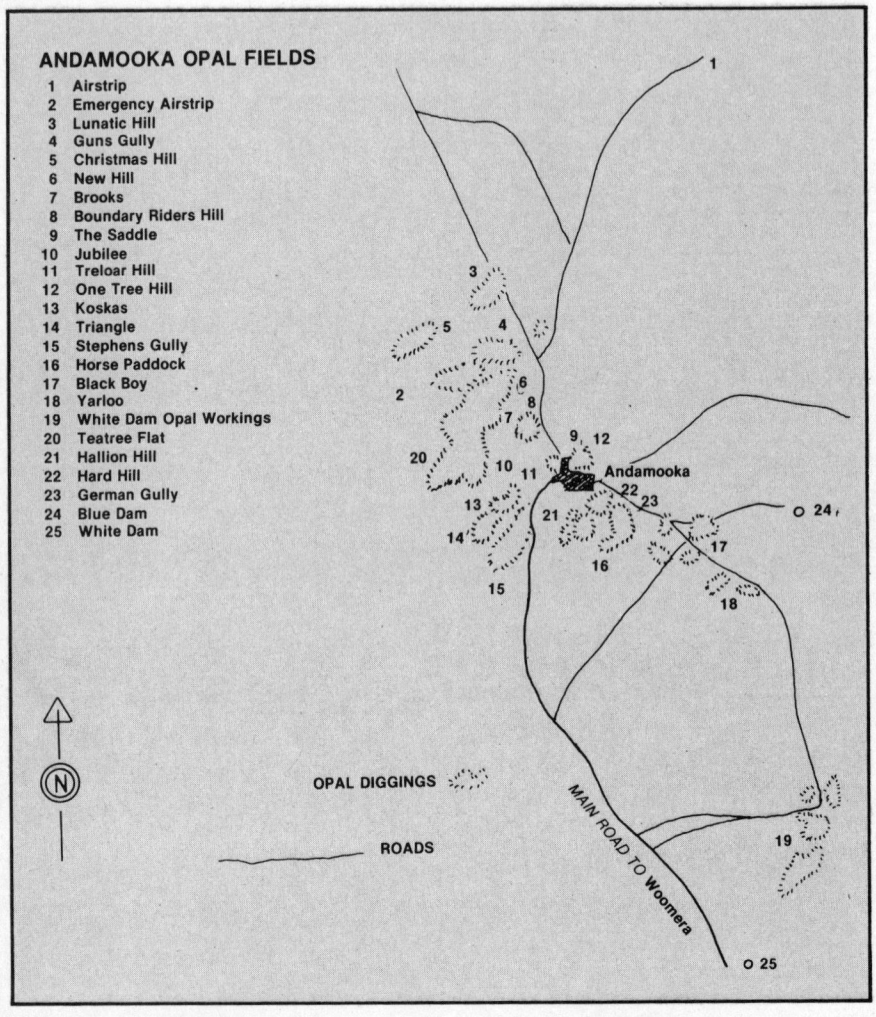

ANDAMOOKA OPAL FIELDS

1. Airstrip
2. Emergency Airstrip
3. Lunatic Hill
4. Guns Gully
5. Christmas Hill
6. New Hill
7. Brooks
8. Boundary Riders Hill
9. The Saddle
10. Jubilee
11. Treloar Hill
12. One Tree Hill
13. Koskas
14. Triangle
15. Stephens Gully
16. Horse Paddock
17. Black Boy
18. Yarloo
19. White Dam Opal Workings
20. Teatree Flat
21. Hallion Hill
22. Hard Hill
23. German Gully
24. Blue Dam
25. White Dam

Andamooka opal fields, South Australia. (Not to scale.)

SOUTH AUSTRALIA

If you wish to search for opal you will need a Precious Stones Prospecting Permit, obtainable from the Department of Mines and Energy or local mining warden. This is valid for one year. With a Permit in your possession you can 'go for your life', but remember to respect other claims.

Access to Andamooka can be by road or air. The town is not on a main highway, but tucked away in the red desert. The road has now been upgraded to a fully sealed state. The Woomera rocket range has also been recommissioned and the Roxby Downs mine has been activated. All this has

An old Andamooka dugout residence. This and several others like it constitute a museum of older mining accommodation.

meant that Andamooka has been rejuvenated in recent times. It now serves as a dormitory town for many of the miners from Roxby Downs. Population stands at approximately one thousand souls, though this number varies from time to time. The people's daily needs are met by a supermarket, 'as comprehensive as any in the city' we are assured. There are no automatic teller machines in Andamooka, but credit cards may be used at the supermarket.

It is a great pity that the people who travel to see the real Australian outback usually miss out on Andamooka. This town is a real outback oasis, with none of the glitter which now attends the larger opal towns. Andamooka has a character all its own, and more people should see and appreciate its old-world charm.

The life Andamooka offers favours more the young person who wishes to make a lucky strike before taking on the responsibilities of a family. But as on most other opal fields there are still the 'old-timers' who have lived at Andamooka for many years, and will never leave. It is a slow, placid, friendly community comprised of many migrants, mainly Yugoslavs, Hungarians and Czechoslovakians, all with a story to tell about the day they found their most memorable opal.

The opal found at Andamooka is amongst the loveliest in the world, but like all opal shy of being found. Searching and mining for opal is hard work by any standards, and luck is not always with the miners.

Andamooka is a community that would welcome an industry, to take up the slack labour. One suggestion put forth is for a pottery, as the surrounding clay has been found suitable for ceramics. In particular, for members of families who are not actively engaged in mining and young people who have left school and need employment.

Because there is never any certainty of opal being found, many families in Andamooka live on a subsistence level.

Some of them go grape picking in the Barossa Valley during the summer months. They are frugal, and when they return to Andamooka they husband their finances carefully to carry them through the winter, should they fail to find opal.

But Andamooka appeals to the people. They find their utopia there, and it is the closest place to paradise they could ever imagine. Civic problems and disturbances are extremely rare, and residents can indulge their individuality and live in perfect freedom. There is just one serpent in this Garden of Eden—there is no future for the children of Andamooka. Unless they become opal miners the young people must leave their homes and parents to find careers far away.

In the meantime, Andamooka sprawls in the sun in its unique atmosphere of 'great expectations', with the realisation of every dream just around the corner.

Local residents of Andamooka are justly proud of their pleasant, peaceful community. Jubilee celebrations were held in 1980 to commemorate the town's fiftieth year. In spite of early setbacks, it now seems certain that as long as the low flat hills continue to yield opal, Andamooka will continue to grow, although perhaps not to the extent that Coober Pedy will grow, as the latter has the distinct advantage of being situated on the highway between Port Augusta and Alice Springs.

Coober Pedy opal fields

ACCESS

Road Access to Coober Pedy has improved since the road was bituminised in 1987; an interesting ten-hour drive is a very welcome change from the erstwhile corrugated horrors of the previous road.
Coach Four coach companies service Coober Pedy.
Air Kendell Airlines runs a daily service from Adelaide.

ACCOMMODATION

Many of the town's residents live underground, and this style of accommodation is also available for tourists. Anything from camping to

A FOSSICKER'S GUIDE TO GEMSTONES

1 Airfield
2 Big Flat
3 Brown's Gully
4 Black Flag
5 Coober Pedy
6 Dead Horse
7 Dead Man's Gully
8 Dora Gully
9 German Gully
10 Geraghty Hill
11 Hans Peak
12 Hospital Hill
13 Ice Cream Hill
14 Jasper Gully
15 Emu
16 Jasper Hole
17 Olympic
18 Potch Gully
19 Shell Patch
20 Stony Hill
21 The Breakaways
22 The Jungle
23 The Companies
24 The Crater
25 Turkey Ridge
26 Willows
27 Yellow Hill
28 East Pacific
29 New Ryans
30 Ryans
31 Greek Gully

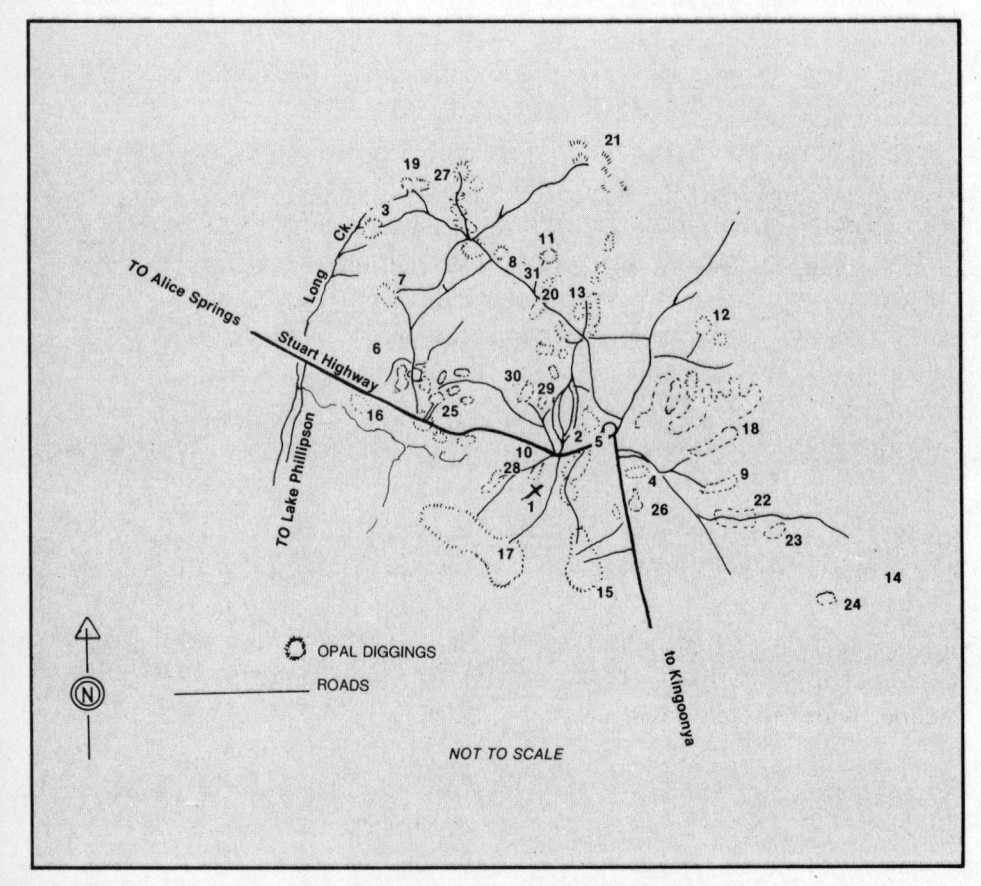

Cooper Pedy opal fields, South Australia.

accommodation in five-star hotels is possible in this well-served centre. Fifteen accommodation houses in all cater for the budget traveller, backpacker, or comfort-seeking group. Two hotels provide further choices. All houses and hotels can arrange tours of the area with experiences to suit the most demanding and fastidious. Every mode of touring is available. The visitor may engage in air or coach tours, safari tours of varying duration, escorted or unescorted walking tours, mine tours or 'noodling' (fossicking) tours. Eating houses and restaurants of every ethnic variety are on tap, with national dishes and experiences not to be missed. As the Coober Pedy Tourist Association says, 'Come and see us, we're different.' The Association has on offer many brochures which are freely available from tourist agents in most states.

Coober Pedy is situated on the north–south Stuart Highway which connects Adelaide in South Australia and Darwin in the Northern Territory. It is 845 kilometres north of Adelaide and 690 kilometres south of Alice Springs.

This world-famous town has grown beyond belief since we first described it in 1980. It seems to be of almost city size and provides all amenities to the opal miners working out of the immediate town within a radius of about forty kilometres. It is in the middle of a siltstone desert which is almost treeless. Very low rainfall (175 millimetres, or 5 inches, the yearly average) combined with high summer temperatures (between 35 and 48 degrees Celsius) inhibits green growth.

The water supply to the town has been a great problem over the years with rationing the accepted norm. A solar still was begun in 1966 but ceased to operate in 1969. Since 1987 a reverse osmosis plant which processes bore water for reticulation has been successfully operated by the District Council. Water is available to tourists at a coin-operated dispenser in Hutchison Street for twenty cents per thirty litres. The council hopes that with the availability of this cheaper water the area will become greener.

An experimental scheme for producing electricity with an air turbine is being trialled by the council in partnership with the Electricity Trust of South Australia and the Office of Energy. The output from the installation is utilised in conjunction with that of the existing diesel generating plant as part of a feasibility study for supplying energy in remote situations.

Noodling is permitted with permission of the owner of the claim, and many of the tours available include fossicking in their itineraries. The council advises all fossickers to be on their guard when walking around claims as very deep shafts exist, some of them abandoned, and these can be traps for the unwary.

It is of note that an opalised pliosaur found in Coober Pedy is on display in the Australian Museum at College Street, Sydney. This member of the fish family is extremely ancient, existing somewhere between 100 and 120 million years ago; it is worthwhile viewing.

The District Council of Coober Pedy estimates that the population now stands at 3500. The majority of citizens are of European origin, most having migrated to Coober Pedy after World War II, and there are about four hundred Aboriginal people living in the town with about one third that number under the age of eighteen years. The town claims to be one of the most ethnic communities in South Australia with forty-five nationalities represented.

Coober Pedy in combination with nearby Andamooka and more distant Mintabie presently produces about ninety per cent in quantity of the world's opal.

Mintabie opal fields

ACCESS

Road Access is via a turn-off 5 kilometres past Marla on the Stuart Highway and 455 kilometres south from Alice Springs. A permit to enter the area costs $5 and is obtainable from the police station. We are warned by staff at the post office that the roads are horrible but trafficable. Enquiry from the auto associations should be sought as usual before starting out.

ACCOMMODATION

The town has a hotel which offers accommodation. There are also two caravan parks to cater for the traveller.

The comparatively unknown field of Mintabie was discovered in 1931; however, it proved too difficult of access and too harsh in climate to be a producing field, given that everything used in mining had to be carried in. The manual methods employed at the time added to the difficulties of the location, and it was not until the advent of heavy machinery in opal mining that strikes were made and a mini-rush came about.

In 1976 the field was rediscovered and fabulous finds were made. Since that time an oversupply of mechanised aid along with the high cost of fuel transport and of maintaining such a force in so remote a location has led to a diminishing in population and perhaps also in the enthusiasm of those first days. But high-quality stone has been mined here and very high prices realised. The population of the town is at present about 300 persons.

The town has two general stores, a hardware store, post office and some accommodation which is outlined above. Reticulated water is available, but being of bore origin we are warned against drinking it. Power is supplied only by individual generators. The main activity in the town is opal mining and at the time of writing in 1996 we are told it is very quiet.

Gemstone occurrences

Algebuckina
  diamond
Andamooka
  precious opal
Angaston
  apatite, common opal, hematite
Arkaroola
  actinolite, azurite, chrysocolla, garnet, malachite, nephrite jade, pyrite
Athelstone
  rock crystal

Balcanoona Station (O'Donahue's Castle)
  malachite
Barossa Range
  topaz
Bimbowrie
  andalusite, apatite, chiastolite, epidote
Brukunga
  pyrite
Burra Burra
  azurite, chrysocolla, malachite

Coober Pedy
  precious opal
Cowell
  hyalite, nephrite jade
Echunga goldfields
  diamond
Ethiudna Mine (Plumbago Station)
  chrysocolla, garnet
Eyre Peninsula
  hematite, nephrite
Iron Knob
  jaspilite
Kangaroo Island
  topaz, tourmaline, zircon
Kanmantoo
  garnet
Kimba
  amethyst
King's Bluff (Olary)
  rock crystal
Melrose
  amethyst, pseudomalachite
Millendilla
  garnet
Mintabie
  precious opal
Montaro
  lazulite
Moonta
  amazonite, amethyst, azurite, fluorite, hematite, malachite, pyrite, smoky quartz, tourmaline
Morgan
  aragonite
Mount Crawford
  sapphire
Mount Davies
  beryl, chrysoprase, peridot
Mount Gambier
  peridot
Mount Gee
  amethyst, carnelian
Mount Howden
  chiastolite
Mount Painter
  garnet, malachite, turquoise, zircon
Mount Pitt
  iolite, ruby, sapphire, tourmaline
Mutooroo
  fluorite, pyrite
Myponga
  apatite
Nairne
  garnet
Olary Province
  actinolite, epidote
Pernatty Lagoon
  azurite, fluorite, malachite
Plumbago Station
  fluorite, garnet, smoky quartz
Port Lincoln
  amazonite
Radium Creek
  sphene
Radium Ridge (Arkaroola)
  carnelian
Strathalbyn
  pyrite
Tassie Creek
  rock crystal
Tourmaline Hill (Arkaroola)
  garnet, tourmaline
Tumby Bay (to Port Lincoln)
  amazonite
Victor Harbour
  andalusite, chiastolite
Wallaroo
  hematite, tourmaline
Williamstown
  beryl, common opal
Wilpena Hill
  beryl, common opal, jaspilite
Yorke Peninsula
  malachite, tourmaline

Northern Territory

THE NORTHERN TERRITORY, affectionately known as the top end, is 1 346 200 square kilometres in area, most of which lies in the tropical and subtropical zones. Much of the Territory is arid, and in many areas desert of the most forbidding kind. Rainfall is sparse and most erratic, and ground temperatures rarely drop below 21°C during daylight hours. Night-time brings the reverse—temperatures as low as several degrees Celsius below zero.

The population of the Territory is concentrated in the coastal north and the southern interior—at Darwin and Alice Springs respectively. There are approximately 98 000 people in total living in these two centres. The coastline is relatively short in relation to the state's vast area, but is exposed to the tropical onshore seasonal winds which greatly influence the climate in the coastal area.

The Government of the Northern Territory looks favourably upon fossicking as long as the fossicker holds a current fossicker's licence. The Department of Mines and Energy has issued an informative free pamphlet for the fossicker entitled *Discover the Undiscovered: Fossicking in the Northern Territory*; this is widely referred to as a 'fossicker's kit'. The pamphlet comprises a comprehensive summary of the law as it affects the amateur fossicker. The contents are edited by a practising fossicker within the Department who has interpreted the law for new and experienced fossickers in a most accessible manner. Various essential knowledge is included.

The Department defines 'fossicking' as a non-commercial recreational activity that involves searching for and collecting rocks, minerals, crystals or fossils by hand or using hand-held implements to a depth of only one metre.

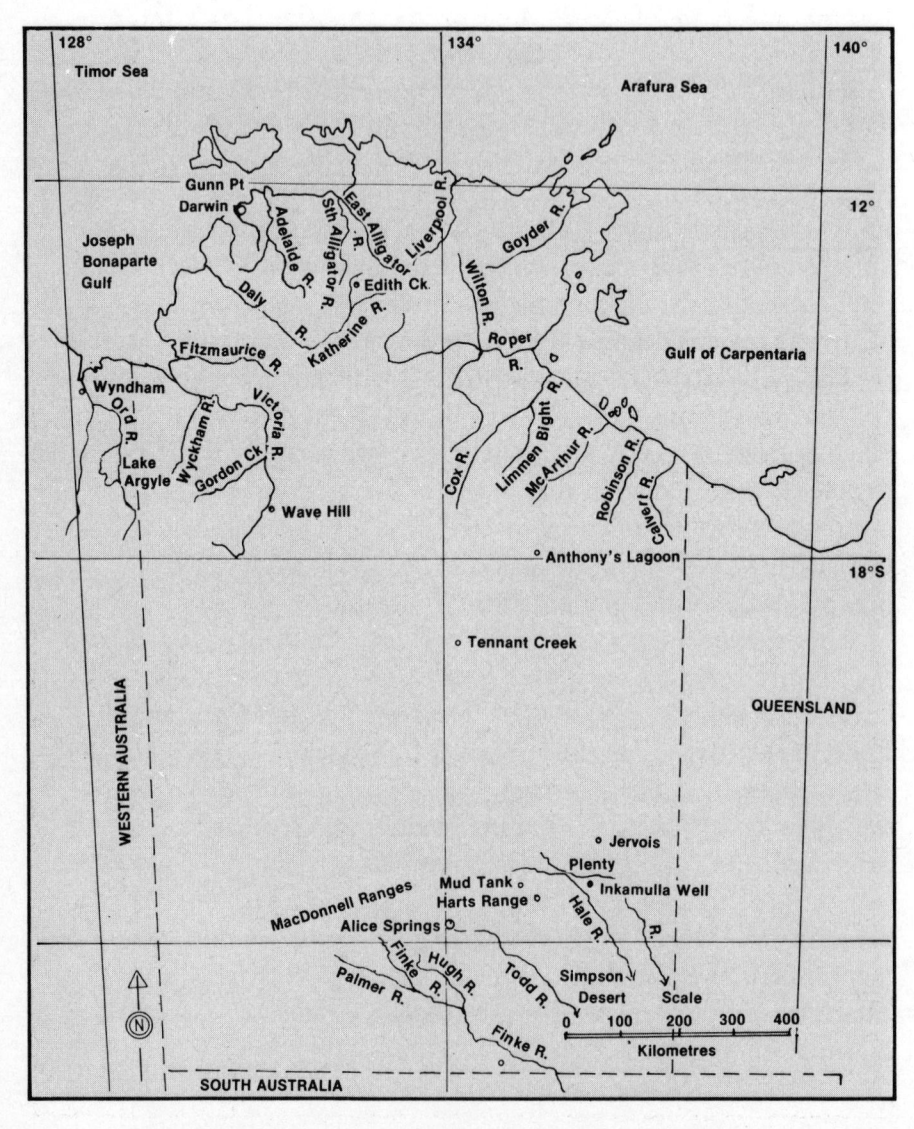

Principal rivers and gemstone-fossicking areas of the Northern Territory.

Discovered treasure can be valuable, perfectly formed, colourful and unusual. The value of any find lies in the eye of the beholder.

The fossicking amendments to the *Mining Act 1996* (NT), which came into effect on 7 May 1996, marked an innovative and exciting new direction for fossicking in the Northern Territory. The *Mining Act* now positively affirms the Northern Territory Government's commitment to promoting fossicking as a recreational and tourist activity. The changes in the legislation distinctly separate the activities of fossicking from those of commercial mining, simplify administrative procedures, and provide fossickers with far greater access to gem and mineral rich areas.

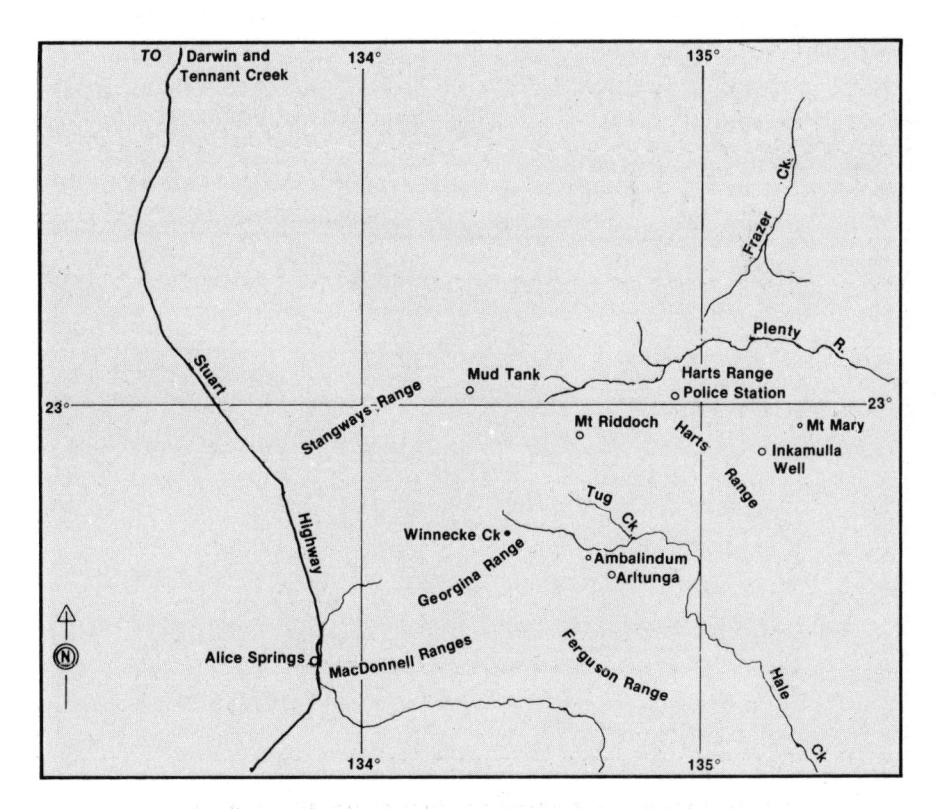

The Alice Springs–Harts Range area of the Northern Territory.

To fossick in the Northern Territory you must hold a Fossicker's Permit issued under the *Mining Act*, section 130 (c). The cost of such a permit is $5 for one month, $10 for two months, $20 for one year and $50 for a five-year period. A Commercial Tour Operator's Licence is also available at $300 for one year. (Note that persons *on* a commercial tour do not require a Fossicker's Permit.)

Hand tools only may be used in fossicking areas—no mechanisation is allowed. The removal of large quantities of any find from a site is not considered as fossicking and the Department emphasises the need for fossickers to show courtesy to one another by taking only a sufficient amount for amateur purposes. The detailed maps available from the Department clearly set out permitted areas and allow quite some latitude for exploration within the designated areas. Under the new Act these are essential reading when planning a fossicking safari. The maps cover ten areas designated for fossicking and are numbered from FA1 to FA11, with FA3 missing. Each of these fossicking areas is described further on in this chapter.

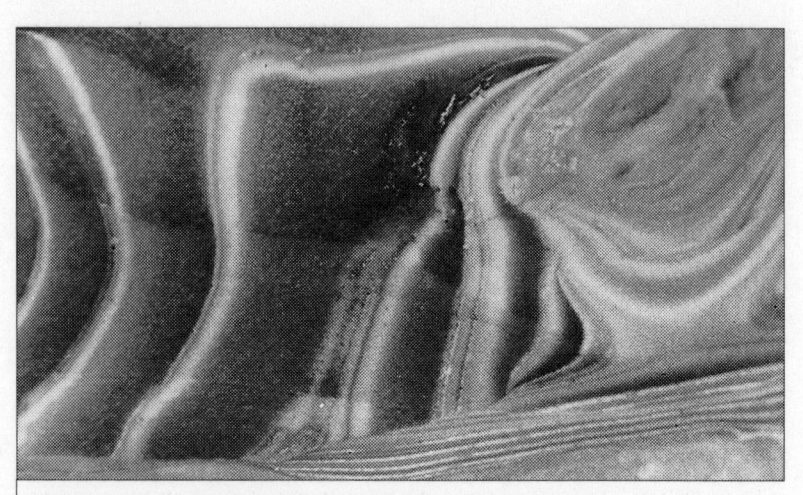

Ribbonstone, formed by cemented sediments from the ancient inland seas of the Northern Territory.

There are two important notes from the Department of which the intending fossicker should be aware of. The first reads:

> Important note for all fossickers. Certain fossicking tour operators, by special agreement with the Department of Mines and Energy, have approval to reserve a number of digging holes between tours by safely placing tools in the holes. Please don't interfere with these reserved holes.

The second note is particularly important:

> Water is unavailable at Mud Tank! You are most welcome to replenish your water supplies with bore water from Gemtree Caravan Park. As a matter of courtesy please notify the office beforehand and collect water only during daylight hours!

The importance of obtaining the correct information on any proposed area first-hand from the Department cannot be overstated. Second-hand instructions and information often suffer in the retelling and can lead to unnecessary and avoidable complications.

Roads and safety

The current state of all roads and access tracks to be traversed needs to be enquired upon immediately before setting out. This information is usually available from the nearest police station or municipal office on the leg to be undertaken, and this very enquiry serves the useful purpose of acquainting responsible persons of the presence of a fossicking party in a particular area. This is very necessary in the case of a remote area. The action is essential if all expectations are to be satisfactorily concluded.

The climate in the Northern Territory is usually very dry, but it is not impossible for short, unseasonal downpours to occur, or in the wet season (summer) for heavy rain to turn the regularly dusty dry but trafficable roads into quagmires in a very short space of time. Such an event can cause very great delay in a schedule, and the notification of authorities of the presence of a party in an area will provide a safety measure. Safety first is a very real and necessary precaution in the Territory—indeed on any gemstone-fossicking safari.

Fossicking Area 1

The Department's Map FA1 shows the Central Harts Range area, approximately 200 kilometres from Alice Springs. Finds of sunstone, garnet, quartz and black tourmaline have been reported here.

Fossicking Area 2

Map FA2 shows Arltunga, approximately 113 kilometres from Alice Springs, with access off the Ross Highway. Gold is the main mineral of interest here.

Fossicking Area 3

There is no designated Map FA3 at time of writing.

Fossicking Area 4

Map FA4 shows Moonlight Rock Hole, 58 kilometres out from Tennant Creek. This is also a gold possibility.

Fossicking Area 5

Map FA5 shows Ilparra Road, 9.5 kilometres southwest of

the Alice via Ilparra Road. This area has yielded amethyst crystal somewhat varying in colour from dark through light to smoky.

Fossicking Area 6

Map FA6 shows Maloney Creek, situated 120 kilometres south of Alice Springs on the Stuart Highway. This area provides a very good opportunity to gain experience of fossil trilobites some of which are as old as 485 million years. The Department advises that the gully on the east side of the road towards Alice Springs has yielded many a good specimen.

Fossicking Area 7

Map FA7 shows Glenarro Bore, which is approximately 170 kilometres from Timber Creek, south of the intersection of the Victoria Highway and the Duncan Highway, near the West Australian border. Here zebrastone and banded jasper are to be found. Zebrastone is good for carving and of great age (approximately 1000 million years old).

Fossicking Area 8

Map FA8 shows Wave Hill, southwest of Katherine by about 340 kilometres, with access off the Lajamanu Road. Prehnite is found here, together with amethyst, jasper, agate and smoky quartz.

Fossicking Area 9

Map FA9 shows the Harts Range West area, approximately 173 kilometres from Alice Springs off the Plenty Highway.

This area is rich in goodies, which include, among others, graphic pegmatite, aquamarine, black-and-green tourmaline, quartz, green epidote and garnet.

Fossicking Area 10

Map FA10 shows Oneva Creek. Apatite, garnet, black magnetite and pink microcline have all been recovered in this area, apatite and garnet being the primary yield.

Fossicking Area 11

Map FA11 shows the Mud Tank zircon field. This is a fine zircon source where apatite, garnet, magnetite and microcline have also been discovered. It is about 147 kilometres from Alice Springs with access off the Plenty Highway.

Gemstone occurrences

Alice Springs (425 kilometres northeast of)
　cassiterite, garnet, turquoise
Anthonys Lagoon
　ribbonstone
Daly River
　azurite
Disputed Mine (Harts Range)
　amethyst
Finke
　australite (tektite)
Harts Range
　amethyst, apatite, beryl, cassiterite, epidote, feldspar, garnet, iolite, kyanite, sapphire, smoky quartz, sphene, tourmaline, zircon
Jervois Range
　azurite, beryl, garnet
MacDonnell Range
　garnet
Mud Tank (Jemkin's Old Camp)
　apatite, zircon
Northern Territory
　buffalo horn
Plenty River
　amethyst, sapphire
Rum Jungle
　azurite
Strangways Range
　zircon
Tennant Creek
　azurite, ribbonstone
Yambah Station
　citrine

Western Australia

ALTHOUGH WESTERN AUSTRALIA has many known gem deposits, and suspected deposits, accessibility in this vast state is one of the big problems for fossickers.

Mining companies have pegged most of the areas with known deposits, and poor roads and hostile climatic conditions are not favourable to amateur fossickers, except perhaps the most dedicated. In certain areas it has also been reported that local residents do not welcome fossickers.

A Miner's Right must be purchased from the head office of the Department of Minerals and Energy or from any mining registrar before setting out on a fossicking safari. The licence is valid for twelve months and a fee of $20 is payable at the time of writing. The possession of this Miner's Right does not bestow the ability to enter private land without the permission of the owner.

Fossicking is defined in the *Mining Act* as searching for, extracting and removing rock, ore or minerals *other than gold or diamonds* not exceeding twenty kilograms for a mineral collection, lapidary work or hobby interest—by use of hand tools only. In this context fossicking is limited to activities associated with the collection of mineral specimen, lapidary work or hobby interests and consequently the use of a metal detector is not permitted. When searching for gold by authority of a Miner's Right a metal detector may be used.

Many gem materials dear to the hearts of amateur lapidaries are found in Western Australia. Agate and jasper

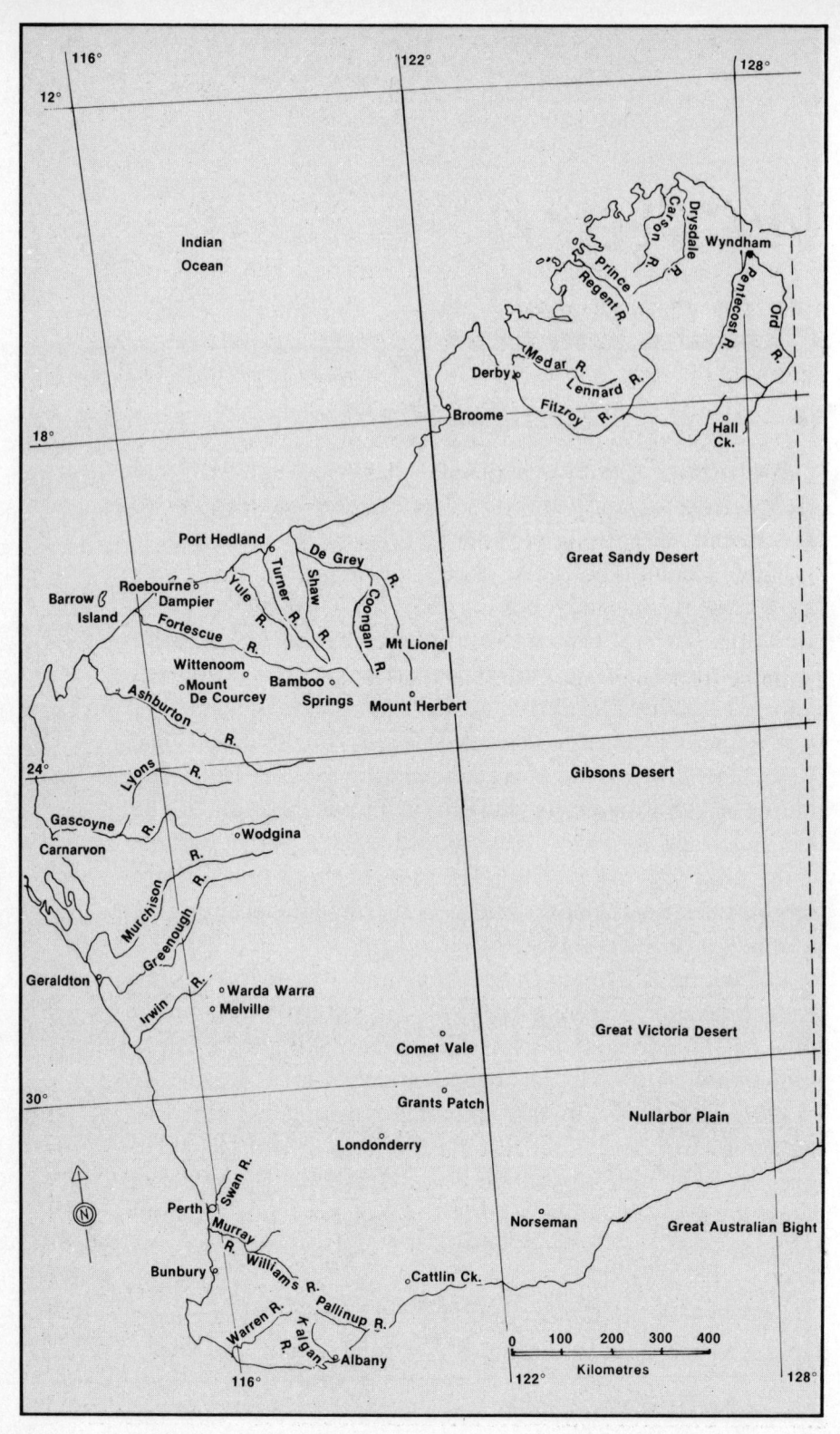

Principal rivers and fossicking centres of Western Australia.

are in plentiful supply. Chrysoprase, tigeriron, opalite, amethyst, ribbonstone, prehnite, chrysocolla, chrysoberyl, emerald and diamond have been found, but the areas of occurrence are widespread.

The areas of gemstone riches usually lie in harsh and unwelcoming terrain, and because of the vast distances to be travelled the cost of fossicking can be quite high.

The head office of the Department or any mining warden's office can supply the fossicker with a number of valuable brochures of great benefit. It is essential, as has been said earlier in this book, that any official information concerning any proposed fossicking venture be collected and understood so that the trip may be as free of problems as possible. The booklet *Gemstones in Western Australia*, available from the Department, is particularly recommended.

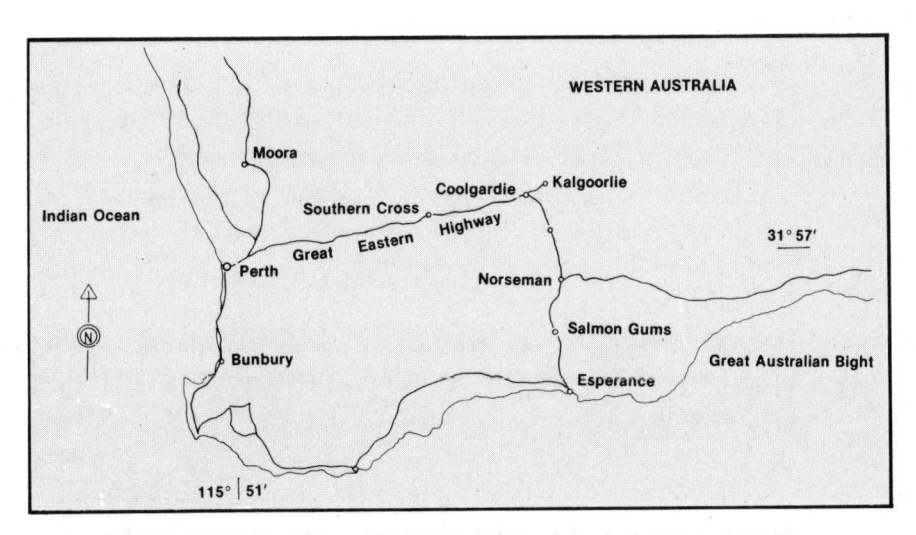

South Western Australian gem fields.

ACCESS

Up-to-date maps of gem areas are obtainable from the Department of Minerals and Energy, and road maps may be obtained from automobile clubs. Always check on the state of the roads before setting out.

ACCOMMODATION

Accommodation en route can be provided by motels, camping grounds and caravan parks. The quality of facilities and service varies greatly, and it behoves travellers to keep an open mind on excellence.

Members of the **beryl family** of gems have been found from time to time in Western Australia, in small quantities, and the lucky fossicker may uncover more. Aquamarine has been found at Yinnietharra in the Gascoyne goldfields and at Spargoville, about fifty kilometres south of Coolgardie. Small amounts of heliodor, the yellow transparent variety of beryl, have been found at Katterup, and a few examples of morganite have been found at Poona. Emerald occurs at Poona, where quite high-grade stones have emerged. But, once again, where worthwhile deposits are found, mining companies have taken out leases.

Precious opal has been found at Coolgardie, as a lining to fissures in a graphitic chlorite schist. Many years ago stones with beautiful fire were obtained from this locality.

Common opal occurs in the Kalgoorlie–Bulong and Coolgardie–Norseman areas, and also at Cue. The type of opal varies with each locality, with lace opal at Bulong, white opal from Bullfinch, flame opal from Mundiwindi and an area about thirty-seven kilometres northeast of Laverton, and red, white and black opal from Widgiemooltha.

Opalised or **petrified wood** is found at the base of the Stirling Range near Cranbrook, Ongerup and Gnowangerup, also at Dandaragan and in the Poole Range in the Kimberley Division.

Some **topaz**, varying from transparent to opaque in pale blue, has been found at Londonderry, Melville and Dalgaranga.

Attractive **jasper** comes from Kalgoorlie, Mt Monger, Marble Bar, Cue and Payne's Find.

Agate is widespread, especially throughout the Antrim Plateau, with Agate Hill, Mt Frank and an area sixty-four kilometres north of Hall's Creek the best known localities. A green-banded variety of agate has been taken from Mt Herbert, and other colours occur at Bamboo Springs and Ilgarari.

Amethyst of varying quality occurs in vesicles and geodes in basalts in the Antrim Plateau, also in the Whim Creek, Ashburton River and Murchison River areas.

Tourmaline has been found at Cattlin Creek near Ravensthorpe, and also at Dalgaranga, north of Yalgoo, and at Spargoville.

Although **corundum** is known to occur at many places, the gem varieties of transparent sapphire and ruby have not been found.

Diamonds have been attracting the attention of big mining companies who have discovered large deposits of stones suitable for industrial purposes and also of gem quality.

Perhaps some lucky fossicker will one day uncover untold quantities of some major gem in this vast territory.

Gemstone occurrences

Agate Hill
 agate
Antrim Plateau
 agate, amethyst
Ashburton River
 amethyst
Bamboo Springs
 agate
Bulong
 chrysoprase, common opal
Byro Station
 corundum
Calverts White Quartz Hill
 emerald
Carnarvon
 opalite
Cattlin Creek
 tourmaline
Comet Vale
 agate, chrysoprase, common opal, prehnite
Cooglegong
 garnet

Coolgardie
    petalite, precious opal, prehnite
Cowarna
    common opal
Cue
    azurite, emerald
Dalgaranga
    topaz, tourmaline
Dangin
    sapphire
East Kimberley region
    prehnite
Five Mile Well
    jade
Gascoyne River area
    mookaite
Gnowangerup
    petrified wood
Grants Patch
    chrysoprase, common opal
Greenbushes
    zircon
Hall's Creek
    agate
Hamersley Ranges
    azurite, hematite, jaspilite,
    rhodonite, ribbonstone, tigereye
Hatters Hill (near Lake King)
    epidote
Horseshoe area (130 kilometres
    north of Meekatharra)
    psilomelane
Ilgarari
    agate
Jacobs Well
    sapphire
Jilbadja
    beryl
Kalgoorlie (south)
    agate, azurite, spodumene
Kennedy Range
    common opal
Kimberley region
    diamond, opalite, prehnite,
    zebra stone
Kununurra region (inaccessible)
    zebra stone
Lake Rebecca
    common opal

Lennard River
    diamond
Lionel
    common opal
Londonderry
    petalite, topaz
Marble Bar
    jasper, Pilbara jade
McPhee's Patch
    emerald, spodumene
Melville
    beryl, emerald, topaz
Menzies (150 kilometres west)
    emerald
Moolyella
    garnet
Moora
    agate, chert
Mount Augustus
    amethyst
Mount Broome
    sapphire
Mount Deception
    agate
Mount De Courcey
    amethyst
Mount Frank
    agate
Mount Herbert
    agate
Mount Hunt
    tourmaline
Mount Newman
    tigereye
Mount Palmer
    prehnite
Mount Tom Price
    tigereye
Mundimindi
    common opal
Mungari
    andalusite, chiastolite
Murchison goldfields
    emerald
Murchison River
    amethyst
Nevoria
    andalusite, chiastolite

Nornalup
  garnet
Norseman
  azurite, common opal
Northampton
  garnet, marcasite
Nullagine
  diamond
Ord Range
  common opal, zebra stone
Paris Mining Centre
  common opal
Payne's Find
  amazonite
Pilbara goldfields
  emerald
Poona
  beryl, common opal, emerald, topaz
Ravensthorpe
  azurite
Richenda River
  sapphire
Roebourne
  rhodonite, ribbonstone, zoisite
Rothsay
  garnet
Sim Creek
  amethyst
Southeastern Western Australia
  australite (tektite)

Southern Cross
  azurite
Spargoville
  agate, beryl, prase, tourmaline
Wandagee
  agate
Warda Warra
  emerald
Weld Range
  jasper
Widgiemooltha
  common opal
Wingellina
  chrysoprase
Wittenoom
  common opal, tigereye
Wodgina
  beryl, emerald, topaz
Yabberup
  garnet
Yampi Sound
  hematite
Yarra Yarra Creek
  common opal
Yerila
  common opal
Yinnietharra
  beryl, garnet, tourmaline
Yundamindra
  common opal

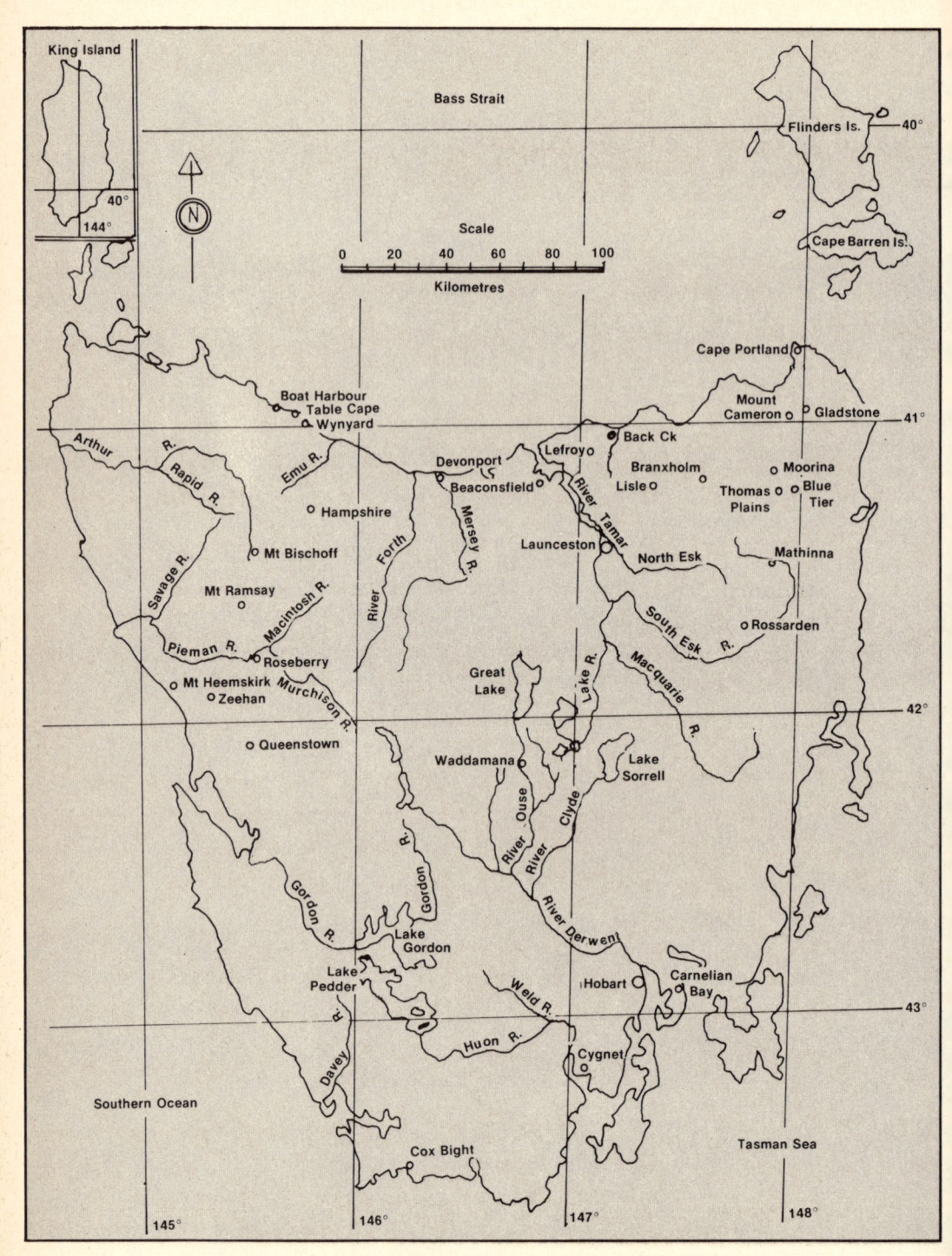

Principal rivers and gemstone areas of Tasmania.

Tasmania

THE ISLAND STATE of Australia lies approximately 65 kilometres off the southeast corner of the mainland. Tasmania may be the smallest in area, 67 800 square kilometres, but it is one of the most heavily mineralised states of Australia. There are many areas of the island which have been extensively mined in the past and subsequently abandoned. In many instances these sites are being reassessed in the light of the present interest in minerals, and the gemstone fossicker can join in this activity, so long as he or she makes a proper investigation of title beforehand from Mineral Resources Tasmania. An addendum to the *Mineral Resources Development Act 1995* was released in November 1995. It is headed 'Prospecting Licences and Fossicking Areas/Division 2—Fossicking Areas'. Item 116(2) states 'Any person may fossick in a Fossicking Area without a licence subject to any conditions the Minister determines.' It is most important to make direct enquiry from the Department before proceeding. Some of the most interesting gem-fossicking areas are set in rather remote areas with difficult access or are on private land. Before setting out on an expedition on private land enquiries should be made from the landholder or lessee, and where necessary permission should be obtained.

At the present time there is no clear indication of firm policy regarding fossicking in World Heritage or Restricted Access areas. Enquiries should be made to the Department of Mineral Resources and to other relevant authorities before making plans to enter such possibly restricted regions. Fossickers should be aware that areas visited on previous expeditions may have since been absorbed into areas designated as national park.

The majority of gemstones found in Tasmania are crystalline or noncrystalline quartz varieties. The precious gemstone varieties are more rarely discovered. However, with modern methods and the intense interest shown by the fossicker today, who is far better off than his counterpart fifty years ago, no one can positively state that no more or other gemstones will be found in a given area. In common with the rest of Australia, Tasmania can never be marked with a stake—we cannot take a map and just pinpoint sites with a confident 'dig here to find such and such'.

Waterworn topaz crystals. These contrast markedly with the previous illustration of topaz (on page 76). Hundreds of years transportation by water and wear by gravels has removed all evidence of the original crystal shape.

The essential research and planning discussed earlier in this book must be carried out to ensure even some possibility of success. A close study of geological formations and the associated minerals found—recently or in the past—could conceivably assist the fossicker in determining the location of undiscovered gemstones. In addition, close liaison with the state authorities, who will be aware of most developments and discoveries in an area, combined with a friendly enquiry from members of the local lapidary organisations will ensure a very good chance of fossicking success.

Accommodation in Tasmania is many-facetted and varied. Fly-drive package tours that include accommodation deals are quite a good proposition for some, while another alternative is to ship a vehicle from the mainland over the strait and camp or stay in hotels, motels or caravan parks on arrival. Hiring a caravan could also be considered. There is quite an amount of planning to be done in regard to this aspect of a trip, but the tourist bureaux in all states are most helpful in this regard.

Many rock shops are owned or operated by super-enthusiastic fossickers who, in most instances, are veritable storehouses of information. These people will dispense their knowledge on request. It is very rare to find a truly keen fossicker unwilling to share his or her experiences of a particular area.

Flinders Island

ACCESS

Air Access to the island may be gained by air. Airlines of Tasmania offer a service from Hobart, from Launceston via Burnie and from King Island. Tamair offers flights from Melbourne and Sydney.

ACCOMMODATION

Accommodation on the island is available at the hotel or guesthouse.

This island is off the Tasmanian coast in the east of Bass

Strait. Tourmaline, topaz and zircon have been found at Killiecrankie Bay, as have petrified wood and fossil shells. The best time for a visit is during the summer months as April to August can be exceptionally cold.

Gemstone occurrences

Anderson's Creek (west of Beaconsfield)
  rhodonite
Arthur River
  zircon
Babel Island
  fluorite
Back Creek
  turquoise
Badger Head
  malachite
Beaconsfield
  amethyst, rhodonite, rose quartz, serpentine, topaz, turquoise, zircon
Bell Mount
  beryl, garnet, topaz
Ben Lomond
  beryl, fluorite
Big Grassy Hill
  amethyst
Birch's Inlet
  stichtite
Blue Tier
  amethyst, apatite, rose quartz
Blythe River
  sapphire, zircon
Boat Harbour
  ruby, sapphire, zircon
Bothwell
  common opal
Branxholm Creek
  sapphire, spinel, topaz
Brown Plains
  topaz
Cameron
  petrified wood
Cape Barren Island
  amethyst, common opal, marcasite, pyrite
Cape Portland
  agate, amethyst, apatite, chalcedony
Cascade River
  malachite
Coles Bay
  cassiterite, sapphire, topaz
Comstock
  garnet
Cornelian Bay
  agate, chalcedony, common opal
Cox Bight
  pyrite, smoky quartz
Cygnet
  common opal, garnet, sphene, spinel
Deloraine
  epidote, peridot
Den Ranges
  turquoise
Doctors Rocks
  peridot
Don Heads
  peridot
Dorset Flats
  topaz
Dugham Range
  common opal
Dundas
  axinite, azurite, crocoite, epidote
East Arm (Tamar River)
  peridot

TASMANIA

Emu River
  amethyst, epidote, garnet, peridot
Flinders Island
  agate, beryl, cassiterite, chalcedony, common opal, smoky quartz, tourmaline, zircon
Forth River (upper reaches)
  peridot
Franford
  malachite
Gawler
  steatite
Gipps Creek
  topaz
Gladstone
  agate, amethyst, sapphire, smoky quartz, spinel, topaz, zircon
Goshen
  cassiterite
Goulds Country
  chalcedony, citrine, common opal
Hampshire
  apatite, axinite, azurite, fluorite, garnet, peridot
Harman Rivulet
  common opal
Heazlewood River
  azurite, garnet, malachite, sphene
Hobart region
  petrified wood
Hudson River
  garnet
Ilfracombe
  agate
Killiecrankie Bay (Flinders Island)
  topaz
King Island
  garnet, spinel
Latrobe
  petrified wood
Launceston
  petrified wood
Lefroy
  amethyst, rose quartz, topaz, turquoise
Lewis River
  garnet
Lindisfarne
  agate, common opal
Lisle
  chalcedony, petrified wood, sapphire
Longford
  petrified wood
Long Island
  zircon
Long Plains
  topaz
Lottah
  fluorite, star sapphire
Lymington
  agate
Macquarie Harbour
  common opal
Magnet Mine
  pyrite, rhodochrosite
Main Creek
  sapphire
Mainwaring Inlet
  azurite, epidote, malachite
Mangalore
  agate
Mathinna
  apatite, topaz
Meredith Range
  chalcedony, zircon
Moina
  beryl, garnet, topaz
Montague
  common opal
Moorina
  amethyst, citrine, rose quartz, sapphire, smoky quartz, spinel, topaz, zircon
Mount Bischoff
  apatite, epidote, fluorite, spinel, tourmaline
Mount Cameron
  agate, amethyst, cassiterite, chalcedony, citrine, common opal, sapphire, smoky quartz, topaz
Mount Claude
  garnet

Mount Heemskirk
  garnet, tourmaline
Mount Kerford (Cape Barren Island)
  garnet
Mount Lyell
  azurite, malachite, pyrite, tourmaline
Mount Montgomery
  tourmaline
Mount Ramsay
  axinite, fluorite, garnet, sphene, tourmaline
Mount Reed
  rhodochrosite
Mount Stewart
  garnet
Mount Stronach
  sapphire
Mount Wellington
  peridot
Northeastern Tasmania
  australite (tektite)
Parkes Hood
  axinite
Parsons Hood
  common opal, sphene
Penguin
  zircon
Pieman River
  chalcedony, common opal, diamond
Proctor's Road
  common opal
Rosebery
  fluorite, rhodochrosite
Rossarden
  amethyst, cassiterite, spinel, topaz, zircon
Ruby Flat
  zircon
Rushy Lagoon
  common opal
Saint Pauls River
  beryl, topaz
Sandy Bay
  common opal
Savage River
  diamond, smoky quartz

Scamander River
  malachite, pyrite
Scottsdale
  peridot
Sea Elephant Point
  garnet
Serpentine Hill
  stichtite
Shag Bay
  common opal
Sidling
  peridot
Sisters Creek
  sapphire, zircon
South Mount Cameron
  amethyst
Southport
  agate
Stanley River
  diamond, sapphire, smoky quartz, topaz, tourmaline
Stoney Ford
  garnet
Supply Creek
  common opal
Table Cape
  epidote, sapphire
Thomas Plains
  sapphire, spinel, topaz, zircon
Thureau's Deep Lead
  emerald
Trial Harbour
  garnet, zircon
Waddamana
  turquoise
Waratah
  cassiterite, peridot
Weldborough
  topaz
Weld River
  alexandrite, ruby, sapphire, topaz, zircon
White River
  garnet
Wilmot River
  peridot
Zeehan
  crocoite, fluorite, malachite, rhodochrosite

Glossary

alluvial deposit Collection of minerals derived from rocks by physical atmospheric action. A concentration of heavier minerals (which would include gemstones) is achieved by the transportation in rivers of the eroded detritus.

beach collecting Fossicking on ocean and lakefront beaches for alluvial minerals. Those found often represent nearby shoreside or subterranean deposits.

contact metamorphism Changes in rocks brought about by heat.

deep lead Sediments which have been concentrated by water and laid down as wash, and which have been subsequently overlaid by new deposits, sometimes more than once. The new deposits may be volcanic outpourings or they may be sections of the Earth's crust that have been windblown, water-washed or physically displaced (for example, as a result of earthquakes). Relocated riverbeds often overlay more ancient versions of the same. Deep leads may also be multiple as the passage of time and geological activity moves layer upon layer.

detritus Gravels produced by the action of water upon rocks.

gemmology The science of gems. Gemmology is just one branch of mineralogy and should be known and understood to a degree if fossicking is to be successful.

geological hammer *or* **pick** A hammer with on one side a striking face which is used to chip mineral samples from boulders or reefs, and on the other side a pointed pick which is used to clear away rubble from a possible specimen rather than to break rock.

hydrothermal precipitation This is the process by which many minerals are brought into being. Typical examples are the quartz family and, of course, opal. This last example differs in that it does not form as crystals but forms a gel.

igneous rock Rock which has formed as the result of the cooling of molten material from the hot internal regions of the Earth. Their composition may vary depending on origin and whether other minerals were contacted while cooling.

lump hammer A short-handled hammer of engineering origin. It has a double-ended head which is comprised of two identical flat striking faces; the head is about one kilogram in weight. The hammer is used primarily for driving gads, chisels and such, but is also useful for breaking smaller pieces of rock.

metamorphic rock Rock which has changed form due to high-pressure contact with other rock or with a new source of hot liquid magma.

mineral A comparatively pure substance which goes to make up rocks. Quartz is a mineral and is composed of silicon in combination with oxygen. Quartz is a form of silica; silica in a mixture with the minerals mica and feldspar comprises what we know as granite.

miner's pick A double-ended pick usually fitted with a short handle for comfortable working in confined spaces.

Miner's Right A written authority obtainable from state mining authorities and their agents for a fee, proving that the holder has permission to fossick on unoccupied crown land. Miner's Rights are not available to persons under eighteen years of age.

nobby A name which is applied to rounded bodies of siliceous material containing opal of varying quality. Nobbies are found in the clay (opal dirt) on the Lightning Ridge opal field in New South Wales.

pegmatite A type of metamorphic rock. Pegmatites form when the rising molten material destined to change the surrounding rocks is high in silica and penetrates into cavities already mineralised with or without crystal formation. The penetration may be into vertical cavities, when the pegmatite formed is referred to as a 'dyke', or it may occur in horizontal cavities and form a 'sill'. Gemstone crystals are frequently found in pegmatites and some may be of considerable size. Fine examples of perfect crystal form may be discovered.

pick mattock A pick which has an adze-like digging tool at one end of its blade. Very good for the rapid loosening of the surface before using the shovel.

primary rock Rock formed by the cooling of original molten magma. Those rocks which erode to produce detritus are known as primary rocks.

regional metamorphism Changes in rocks brought about by pressure.
sediment The material formed as the result of the break-up of rocks by erosion. Categories into which sediments may fall are rubble, gravel, sand and clay.
secondary rock Rock formed when detritus is reconstituted by heat, pressure or cementation.
sedimentary rock The rock formed when sediment is reconstituted by pressure or cementation.
sieve A screen available in various opening sizes used to separate gravels and finely divided minerals. In past times sieves were identified by mesh size, that is by how many openings and wires there were to the inch in the case of sieves below a certain size. Since the advent of the metric system all sieves are known by the actual size of the particle which will fit through a single mesh opening. Three-millimetre and six-millimetre are the sizes of sieve most used by fossickers. A sieve is made by covering a frame of wood or metal with a wire mesh cloth of the required hole size. Sieves also come made from a metal sheet perforated with round holes of suitable diameter. It might be in order to warn fossickers here of the great weight of sieves above thirty centimetres in diameter when loaded with gravel. It takes quite an athlete to manipulate larger sizes of sieve for any length of time.
vug *or* **vugh** A cavity in rock. Vugs are probably formed as the result of a gas bubble being trapped in molten material. As the molten material cools and solidifies condensation of the contents occurs and crystals form on the interior surface. Any subsequent penetration of the cavity by other liquids may add to the growth already present. A vug is a very good prospect for the fossicker.
wash This consists of gravels and sediments which are the result of erosion acting on rocks. Wash is often transported by water and concentrated into pockets and stream bed layers which can form deep leads.
willoughbee *or* **willobie** An ingenious device which comprises a cradle that accepts a sieve and which has a handle fitted in such a way as to allow rapid up-and-down movements of the sieve into a water tank. The willoughbee is used to separate gem gravel from clay.

Further reading

The following publications are suggested as useful additional reading.

Buchester, K. J., *Popular Prospecting*, Rigby, Adelaide, 1977.
Dana and Hurlbut, *Dana's Minerals and How to Study Them*, Wiley, New York, 1963.
Fossicking Areas in New South Wales, New South Wales Mineral Resources.
Gemstones in Western Australia, West Australian Mining Information Centre, East Perth, 1994 (52 pages).
Information Series No. 1—Mining Act 1978 Basic Provisions, West Australia Department of Minerals and Energy, East Perth.
Information Series No. 2—(amended) Miner's Right, West Australia Department of Minerals and Energy, East Perth, October 1996.
Information Series No. 4—Private Land Provisions, West Australia Department of Minerals and Energy, East Perth, 1958.
Information Sheet No. 102, Department of Minerals and Resources, St Leonards, New South Wales, 1992.
Minefact, New South Wales Mineral Resources Information Branch (information sheets).
Opals in New South Wales, New South Wales Mineral Resources/Australian Gem Industry Association.
Perry, N. and R., *Australian Gemstones in Colour*, Reed, Sydney, 1967.
Perry, N. and R., *Australian Opals in Colour*, Reed, Sydney, 1969.
Perry, N. and R., *Gemstones in Australia*, Reed, Sydney, 1979.
Perry, N. and R., *Practical Gemcutting*, Reed, Sydney, 1980.
Sapphires in New South Wales, New South Wales Mineral Resources Information Branch (colour booklet).
Prospectors' Guide to Gold in Australia, Reed, Sydney, 1982.
Webster, R., *The Gemmologist's Compendium*, NAG Press, London, 1970.

Helpful information

The following collection of publications and their source is supplied with the hope that they, aided by the helpful officers of the organisations issuing them, may be of as much assistance to the intending fossicker as they have been to us in our research. (Please note that at the time of publication, these details were correct, however the pamphlets available and government departments in charge of fossicking in each state may change.)

Northern Territory
Fossicking in the Northern Territory Fossicking Kit
4 pp folder with detailed maps

For more information and to obtain the above, contact:
Northern Territory Department of Mines and Energy
GPO Box 2901, Darwin NT 0801
Phone: (08) 8999 5286

Western Australia
Gemstones in Western Australia
52 pp booklet
Prospecting the Right Way
Folder containing a copy of Information Series No. 5 and Miner's Right
Information Series No. 1
8 pp pamphlet

For more information and to obtain the above, contact:
Mineral Titles Division
Department of Minerals and Energy
100 Plain Street, East Perth WA 6004
Telephone: (09) 222 3118

Victoria
Experience Victoria (1996)
465 pp colour book

Available from RACV Offices throughout Victoria
Royal Automobile Club of Victoria (RACV) Ltd
550 Princes Highway, Noble Park Victoria 3174

Most commonly asked questions on Miners' Rights
7 pp photocopy

For more information and to obtain the above, contact:
Department of Natural Resources and Environment
PO Box 41, East Melbourne Vic 3002
Telephone: (03) 9651 7011

South Australia
Flinders Ranges and Outback
Visitors Guide, 76 pp colour book
South Australian Shores
212 pp colour book
South Australian Getaways
108 pp colour book

For more information and to obtain the above, contact:
South Australian Tourism Commission
GPO Box 1972 Adelaide South
Australia 5001   phone 08 8303 2222

HELPFUL INFORMATION

Tasmania
Fossicking Areas
18 pp general information and detailed maps b/w
Extract from Mineral Resources Development Act 1995
Prospecting Licences and Fossicking Areas, Part 5
3 pp leaflet
Sheet titled 'Conditions Relating to Prospecting Licences'
Application form for Prospecting Licence

For more information and to obtain the above, contact:
Department of Development & Resources
PO Box 56, Rosney Park Tasmania 7018
Telephone: (03) 6233 8329

Queensland
Fossicking in Queensland
4 pp colour brochure
Guide to Queensland Fossicking Legislation
4 pp b/w brochure

For more information and to obtain the above, contact:
Queensland Minerals and Energy
GPO Box 2564, Brisbane Qld 400
Telephone: (07) 3237 1659

New South Wales
Location Diagrams of Fossicking Areas in NSW
34 pp photocopy
List of Mineral Publications, November 1995
34 pp photocopy
Sapphires in New South Wales
16 pp colour booklet
Opals in New South Wales
8 pp colour foldout
Gold in New South Wales
15 pp colour booklet

Minefact No. 20, Access to land for Mineral exploration, May 1994
Minefact No. 5, How to Fossick in New South Wales, Aug. 1994
Minefact No. 9 Opal - The Colourful Outback Gem, Aug 1994
Minefact No. 208 Gemstones, June 1995
Minefact No. 209 Gold, June 1995
Minefact No. 55 Araluen Gold, July 1996
Minefact No. 65 Where to find Gems in New South Wales, August 1996
Gold – Some Notes for Fossickers Information Sheet No 102

For more information and to obtain the above, contact:
Dept of Mineral Resources Information & Customer Service
PO Box 536, St. Leonards NSW 2065
Telephone: (02) 9901 8888

The True Story of White Cliffs
1 pp b/w
Broken Hill New South Wales Australia
42 pp colour book, maps etc

To obtain the above, contact:
White Cliffs Opal Fields Tourist Association Inc.
White Cliffs, New South Wales 2836 Australia

South Australia
General Profile of Coober Pedy
8 pp b/w
Coober Pedy Accommodation Directory (to Mar 1997)
21 pp b/w

To obtain the above, contact:
District Council of Coober Pedy
PO Box 265, Coober Pedy South Australia 5723
Telephone: (08) 8672 5298

Walgett Shire and the Lightning Ridge Opal Fields
25 pp colour booklet
Walgett Shire Tours
12 pp b/w maps and information

To obtain the above, contact:
Lightning Ridge Tourist Information Centre
PO Box 1779, Lightning Ridge NSW 2834
Telephone: (068) 290 565

Index

accommodation
  Agate Creek 100
  Anakie gem fields 102
  Andamooka opal fields 121
  Beechworth 113
  Big Bessie 102
  Buchan district 113
  Castlemaine 115
  Chincilla 106
  Coober Pedy 125
  Flinders Island 149
  Graves Hill 102
  Heathcote 114
  Lightning Ridge 80-1
  Middle Ridge 103
  Mintabie 128
  Mount Surprise 104
  O'Briens Creek 104
  New England district 73
  Tomahawk Creek 103
  Western Australia 142
  White Cliffs 85
  Yowah opal fields 105
actinolite 33, 45
agate 33, 45-6, 143
alluvial deposits 27
alunite 33, 46
amazonite 33, 46
amethyst 33, 46-7, 143
andalusite 33, 47
Anakie gem field 14, 29
Andamooka 121-5
apatite 34, 47
aquamarine 34, 47
aragonite 34, 47
Arkaroola 119
australite (tektite) 34, 37, 47
axinite 34, 48
azurite 34, 48

beach collecting 29-31
Beechworth district 113
beryl 34, 48, 142
Big Bessie gem field 101-2
Buchan district 113

cairngorm 34, 49
carnelian 34, 49
cassiterite 34, 49
Castlemaine 115
chalcedony 34, 49-50
chert 34
chiastolite 50

Chinchilla 106
chrysoberyl 34, 50
chrysocolla 34, 50
chrysocolla in quartz 34
chrysoprase 34, 50
citrine 34, 50
climate 5
common opal 35, 51-2, 142
contact metamorphism 10
Coober Pedy 125-8
corundum 143
crocoite 35, 52

deep lead 14
diamond 35, 52, 143
distance 4

emerald 35, 53
epidote 35, 53
erosion 13

feldspar 53
Flinders Island 149-50
fluorite 35, 53
fossicking
  etiquette 22
  manners 16
  methods 25-9
  regulations 19-22, 69, 94-7, 111, 131-4, 139, 147

garnet 35, 54-5
gemmology 15
gemstone
  occurrences
    NSW 86-92
    NT 138
    QLD 107-10
    SA 129-30
    Tas 150-2
    Vic 115-18
    WA 143-5
  principal areas 42-3
  recognition of 33-8
  in various rock groups 10-15
geography 4-6
geology 7-15, 39-3
Glen Innes 76-80
Glenalva gem field 102
goshenite 35, 55
granitic rocks 10
Graves Hill gem field 102
Great Artesian Basin 40-1, 93

INDEX

Heathcote 114
helidor 35, 55
hematite 35, 55
horn 35, 56
hyalite 35, 56
hydrothermal precipitation 12

igneous rock 8
intrusive rocks 10
iolite 35, 56

jade 56
jasper 35, 56, 143
jaspilite 35, 56
jet 35, 56

kyanite 35, 57

labradorite 35, 57
lazulite 35, 57
Lightning Ridge 80-5
lump hammer 16

malachite 35, 57
maps 5-6, 42-3, 70-2, 77, 83, 96-7, 101, 103, 104, 105, 112, 120, 121, 122, 126, 132, 133, 140, 141
marcasite 35, 57
metamorphic rocks 9, 10-11
method 15-18
Middle Ridge gem field 103
minerals 10
Mintabie 128-9
mookaite 35, 57
moonstone 36, 57
morganite 36, 58
morion 36, 58

nephrite 36
nephrite jade 56
New England district 73-80
New South Wales 69-92
Northern Territory 131-8

O'Briens Creek gem field 104
obsidian 36, 58
onyx 58
opal 36
  bearing areas 40, 121, 142
opalised wood 36, 58, 142
opalite 36, 58

'peanut' jasper 36, 58
pegmatites 12
peridot 36, 58
petalite 59
petrified wood 59, 142
Pilbara jade 56, 59
prase 36, 59
precious opal 59-60, 142

prehnite 36, 60
primary rocks 8
pseudomalachite 36, 60
pseudophite 36
psilomelane 36, 60
pyrite 36, 60

Queensland 93-110

regional metamorphism 10
rhodochrosite 36, 60
rhodonite 36, 60-1
ribbonstone 37, 61
rock crystal 37, 61
rocks 8-10
rose quartz 37, 61
ruby 37, 61-2
rutilated quartz 37, 62

safety 22-4, 99, 135-6
sapphire 37, 62-3
secondary rocks 9
sedimentary rocks 9
serpentine 63
sieves 26, 27, 29
smoky quartz 37, 63
South Australia 119-30
sphene (titanite) 37, 63
spinel 37, 64
soapstone *see* steatite
spodumene 64
steatite 37, 64
stichtite 37, 64

Tasmania 146-52
tektites 34, 37, 47
tertiary rocks 9
thunder eggs 37, 64
tigereye 37, 64
Tomahawk gem field 103
tools 16-18
topaz 37, 65-6, 142
tourmaline 37, 66-7, 143
  in quartz 37, 67
turquoise 38, 67

variscite 38, 67
Victoria 111-18

wash 26, 28
Western Australia 139-45
White Cliffs 85-6
willoughbee (or willobie) 29

Yowah gem field 104-5

zebra stone 38, 67
zircon 38, 68
zoisite 68